Project Management Capability Assessment

Internal Audit and IT Audit

Series Editor: Dan Swanson

Cognitive Hack: The New Battleground in Cybersecurity ... the Human Mind
James Bone
ISBN 978-1-4987-4981-7

The Complete Guide to Cybersecurity Risks and Controls
Anne Kohnke, Dan Shoemaker, and Ken E. Sigler
ISBN 978-1-4987-4054-8

Corporate Defense and the Value Preservation Imperative: Bulletproof Your Corporate Defense Program
Sean Lyons
ISBN 978-1-4987-4228-3

Data Analytics for Internal Auditors
Richard E. Cascarino
ISBN 978-1-4987-3714-2

Ethics and the Internal Auditor's Political Dilemma: Tools and Techniques to Evaluate a Company's Ethical Culture
Lynn Fountain
ISBN 978-1-4987-6780-4

A Guide to the National Initiative for Cybersecurity Education (NICE) Cybersecurity Workforce Framework (2.0)
Dan Shoemaker, Anne Kohnke, and Ken Sigler
ISBN 978-1-4987-3996-2

Implementing Cybersecurity: A Guide to the National Institute of Standards and Technology Risk Management Framework
Anne Kohnke, Ken Sigler, and Dan Shoemaker
ISBN 978-1-4987-8514-3

Internal Audit Practice from A to Z
Patrick Onwura Nzechukwu
ISBN 978-1-4987-4205-4

Leading the Internal Audit Function
Lynn Fountain
ISBN 978-1-4987-3042-6

Mastering the Five Tiers of Audit Competency: The Essence of Effective Auditing
Ann Butera
ISBN 978-1-4987-3849-1

Operational Assessment of IT
Steve Katzman
ISBN 978-1-4987-3768-5

Operational Auditing: Principles and Techniques for a Changing World
Hernan Murdock
ISBN 978-1-4987-4639-7

Practitioner's Guide to Business Impact Analysis
Priti Sikdar
ISBN 978-1-4987-5066-0

Project Management Capability Assessment: Performing ISO 33000-Based Capability Assessments of Project Management
Peter T. Davis and Barry D. Lewis
ISBN 978-1-138-29852-1

Securing an IT Organization through Governance, Risk Management, and Audit
Ken E. Sigler and James L. Rainey, III
ISBN 978-1-4987-3731-9

Security and Auditing of Smart Devices: Managing Proliferation of Confidential Data on Corporate and BYOD Devices
Sajay Rai, Philip Chukwuma, and Richard Cozart
ISBN 978-1-4987-3883-5

Software Quality Assurance: Integrating Testing, Security, and Audit
Abu Sayed Mahfuz
ISBN 978-1-4987-3553-7

Supply Chain Risk Management: Applying Secure Acquisition Principles to Ensure a Trusted Technology Product
Ken Sigler, Dan Shoemaker, and Anne Kohnke
ISBN 978-1-4987-3553-7

Why CISOs Fail: The Missing Link in Security Management—and How to Fix It
Barak Engel
ISBN 978-1-138-19789-3

Project Management Capability Assessment

Performing ISO 33000-Based Capability Assessments of Project Management

Peter T. Davis
Barry D. Lewis

CRC Press is an imprint of the
Taylor & Francis Group, an **informa** business
AN AUERBACH BOOK

CRC Press
Taylor & Francis Group
6000 Broken Sound Parkway NW, Suite 300
Boca Raton, FL 33487-2742

Version Date: 20160826

International Standard Book Number-13: 978-1-138-29852-1 (Paperback)

Visit the Taylor & Francis Web site at
http://www.taylorandfrancis.com

and the CRC Press Web site at
http://www.crcpress.com

To all the project managers who had to suffer with me on
their project teams and to all my illustrious forefathers and
foremothers—especially Finlay and Ruth. I know I was a project.

—Peter

To my long-suffering wife, thank you for
everything you do sweetheart!

—Barry

Contents

Part II Process Assessment Method

Foreword

There is growing recognition that organizations are officially doing more "projects." With the growing size and complexity of those projects, more is at stake and performance is far more visible. This creates a need for better, more formalized project management (PM) practices—for both project managers and project management offices (PMOs). Whether you're launching a new PM program, or looking to take your PMO to the next level, take the opportunity to pause and reflect by asking yourself, "Have you implemented the *right* PM processes, the *right way* to improve your chances of success?"

The answer to this question lies in an *assessment* of your, preferably, *evidence-based* practices. With a unique blending and merging of widely-adopted standards, guidelines, and expertise, this PM assessment approach helps to determine *where you are* and *where you need to go*. Barry and Peter have put together an approach to help confirm and accelerate the way forward.

This first-time amalgamation of an organization's most common, leading practices in a single method delivers an eloquent approach to measuring a continuous improvement of PM. You will use evaluation to "score" your current state and validate your organization's ability to meet its PM needs and make an honest assessment of any gaps that need filling on the road to advanced PM maturity.

After this first step, you should develop your roadmap with clear, near- and long-term strategic visibility. Using this capability assessment as a basis, your PM initiatives can then be specified, prioritized, and sequenced for continuing program improvement.

This method is suitable for a wide-range of organizations and PM programs, regardless of the variables listed below:

- Size and level of PM program sophistication
- Stage of development, or maturity of PMO
- Project type or focus

The assessment methods described in this book will allow you to set expectations then graduate your PM Program to the next level.

Steve Tower
P.Eng. CMC PMP MBCI

Acknowledgments

We would like to acknowledge Dan Swanson for pitching the book to the editorial committee and getting us a contract–much appreciated, Dan; Rich O'Hanley, who started this rolling; our copyeditor who kept us honest; and Steve Tower, technical editor, for his diligence in reviewing the material.

Peter would like to thank first and foremost his co-author for taking this trip with him. He would also like to thank Dan Swanson, Steve Tower, Ronn Faigen, and Ivo Haren for listening to his half-baked ideas. The information they provided shows in this book. Any mistakes, as they say, are mine and not theirs.

Barry would like to thank Peter, without whom this book would not have been possible. He also thanks Dan Swanson, Steve Tower, and Ronn Faigen for their help in getting this project off the ground and completed.

Authors

Peter T. Davis (Certified ISO 9001 Foundation, Certified ISO 20000 Lead Auditor/Implementer, Certified ISO 22301 Foundation, Certified ISO 27001 Lead Auditor/Implementer, Certified ISO 27005 Risk Manager, Certified ISO 27032 Lead Cybersecurity Manager, Certified ISO 28000 Foundation, Certified ISO 30301 Lead Auditor, Certified ISO 31000 Risk Manager, CISA, CISM, CISSP, CGEIT, CPA, CMA, CMC, COBIT 5 FC, Certified COBIT Assessor, COBIT Assessor Certificate, COBIT Implementer Certificate, DevOps FC, ISTQB Certified Tester Foundation Level (CTFL), ITIL 2011 FC, Lean IT Association Foundation, Open FAIR FC, PMI-RMP, PMP, PRINCE2 FC, RESILIA FC, Scrum Fundamentals Certified, SSGB) founded Peter Davis+Associates as an information technology (IT) governance firm specializing in the security, audit, and control of information. A battle-scarred information systems veteran, his career includes positions as programmer, systems analyst, security administrator, security planner, information systems auditor, and consultant. He also is the past President and founder of the Toronto Information Systems Security Association (ISSA) chapter, past Recording Secretary of the ISSA's International Board and past Computer Security Institute Advisory Committee member. He has written or co-written numerous articles and 12 books, including *Lean Six Sigma Secrets for the CIO*. He was listed in the *International Who's Who of Professionals.* In addition, he was the third Editor in the three-decade history of *EDPACS*, a security, audit and control publication. He lives in Toronto, Ontario.

Barry D. Lewis held the CISSP, CISM, CRISC, and CGEIT designations until December 2016 when he retired. Prior to that he worked in IT for almost two decades in the banking industry before becoming a consultant in 1987 and starting his own firm, Cerberus ISC Inc., with two partners in 1993. He has presented international seminars on information security and governance for the last three decades on five continents (and is still wishing to fill in the sixth in South America). He won the prestigious John Kuyers Best Speaker/ Conference Contributor Award in 2008 from ISACA, and, in June of 2017, he won the Bob Darlington Best Speaker and, Friend of the Chapter award from the Toronto ISACA chapter. He has authored numerous papers over the years and co-authored over a half-dozen books with Peter, the last being one of the signature Dummies series called *Wireless Networks for Dummies.* Barry lives in Burlington, Ontario, with his wife of 37 years. They have one son and a ragdoll named CK.

Reviewer

Steve Tower is an IT professional and experienced management consultant with several years as a technology leader. He specializes in business applications, IT-related process improvement, and information risk management. He was the CIO of a professional services firm and the practice director of three top-tier consulting companies. Steve has led complex application software, process improvement, and infrastructure projects for international, Fortune 100 companies. Steve has been teaching Project Management Professional (PMP®) exam preparatory courses for several years and is a frequent expert contributor to independent research on IT disaster recovery. He is a Professional Engineer (P.Eng.), Certified Management Consultant (CMC), Project Management Professional (PMP), and a member of the Business Continuity Institute (MBCI).

Why Should I Buy This Book?

This book is the convergence of three great International Organization for Standardization (ISO) standards: ISO 21500, ISO 33020, and ISO 33063. You will find a process assessment model or method based on ISO 21500:2012 that is compliant with the ISO/International Electrotechnical Commission (IEC) standard ISO 33020:2016. This comprises Part I of the book. The model using ISO 21500 forms the basis for assessing the capability of the processes in a project management system based on ISO 21500.

In Part II, you will find an assessor guide that illustrates how to undertake an assessment based on the ISO 33063 standard. This process is evidence-based and will give the assessor—and the assessed—a reliable, consistent, and repeatable assessment process in the governance and management of project management. Using a standard based on ISO 33020 will help project managers and directors gain executive and board member buy-in for change and improvement initiatives. You can find a list of the relevant ISO/IEC 33000 Process Assessment Standards in the section of this book titled "What are the ISO 33000 Standards?"

The assessment model serves two different purposes. First, it will help assist with process capability determination. Second, it will assist with process improvement.

In Part I, you will find a process reference model and a set of process indicators of process performance and process capability that you will use in Part II as the basis for collecting objective evidence during the assessment so that you may assign a capability rating. You may use that rating as a target or as the basis for improvement.

This book offers:

- The only widely available assessment method that provides an enterprise-level view of project management process capability, providing an end-to-end business view of project management's ability to create business value
- Developed and based on the deep knowledge and experience of ISO, a widely recognized global leader
- Enables assessments by enterprises and skilled assessors to support process improvement

Why Do We Need This Method?

One of the authors was in a project management LinkedIn group, when someone made a revelatory statement—at least to the author. The group member stated that the Dutch government was interested in creating something that would help them forecast the success of private-public partnerships. These *public–private partnerships* (PPP, 3P, or P3) are a cooperative arrangement between two or more *public* and *private* sectors, typically a project of a long-term nature. So, we are trying to forecast the distant future. Now, not many of us have a crystal ball—and should we, we might not use it for something as mundane or altruistic as looking at the success of government programs. But this is a classic problem associated with waterfall project management methodologies: you fix the date instead of the deliverable. This causes two problems. One, Parkinson's Law states that "work expands so as to fill the time available for its completion." So, build enough slack into your projects and you are guaranteed to waste precious resources. The second problem is that dates are often fixed without sufficient knowledge of portending problems. Consequently, quality must suffer to meet artificial or capricious deadlines. This kills morale or allows a false-positive morale from finally getting the project completed, making stakeholders unhappy. Not to say anything about the poor products.

However, if there was a way to assist you in forecasting project success, would you use it? The question led to a revelation by one of the authors—you could assess the capability of every process in project management *and* arrive at organizational maturity for project management. The more mature the organization, the lower the risk associated with project management. Logically, when you have the best inputs and the best process, you must have the best output. So, should you want to forecast future project success, you should look at process capability. While not infallible—humans are involved in the processes and the assessments—it is worthwhile to assess your processes and work on closing gaps.

We could regale you with a litany of world-class project failures—the Internal Revenue Service's (IRS) Business Systems Modernization, U.S. National Reconnaissance Office Future Imagery Architecture, U.S. Federal Aviation Administration Advanced Automation System, Kmart IT systems modernization,[1] et al. But that is not the purpose of this book. The purpose of this book is to help your organization from adding its name to the list.

We did not say we would not provide some shocking project management statistics. Did you know that:

1. Up to 75 percent of business and IT executives anticipate their software projects will fail. (From https://www.geneca.com/blog/software-project-failure-business-development.)
2. Fewer than a third of all projects were successfully completed on time and on budget during 2013. (From https://www.versionone.com/assets/img/files/CHAOSManifesto2013.pdf.)
3. Seventy-five percent of IT executives believe their projects are "doomed from the start." (From https://www.geneca.com/blog/software-project-failure-business-development.)
4. Over 1 in 3 (about 34%) projects have no project baseline. (From http://www.wellingtone.co.uk/wp-content/uploads/2016/01/The-State-of-Project-Management-Survey-2016.pdf.)

[1] The first clue from the list is don't "modernize"—as this is equivalent to "putting lipstick on a pig" or "rearranging the deckchairs on the Titanic"—but innovate. This is the rationale behind Level 5 of the model.

5. For every $1 billion invested in the United States, $97 million was wasted due to poor project performance. (From http://www.pmi.org/-/media/pmi/documents/public/pdf/learning/thought-leadership/pulse/pulse-of-the-profession-2017.pdf.).
6. Fifty percent of all Project Management Offices (PMOs) close within just three years. (From https://www.apm.org.uk.)
7. An astounding 83 percent of senior executives fully understand the value of project management to the business. (From https://www.pmi.org/-/media/pmi/documents/public/pdf/learning/thought-leadership/pulse/pulse-of-the-profession-2017.pdf.).
8. Eighty percent of project management executives don't know how their projects align with their company's business strategy. (From https://www.changepoint.com/resources/articles/survey-did-you-know/.)
9. High-performing organizations successfully meet original goals/business intent in 92 percent of their projects, while low performers complete only 3 percent. (From https://www.pmi.org/-/media/pmi/documents/public/pdf/learning/thought-leadership/pulse/pulse-of-the-profession-2017.pdf.).
10. Seventeen percent of large IT projects go so badly that they can threaten the very existence of a company. (From http://www.mckinsey.com/business-functions/digital-mckinsey/our-insights/delivering-large-scale-it-projects-on-time-on-budget-and-on-value.)
11. On average, large IT projects run 45 percent over budget and 7 percent over time, while delivering 56 percent less value than promised. (From http://www.mckinsey.com/business-functions/digital-mckinsey/our-insights/delivering-large-scale-it-projects-on-time-on-budget-and-on-value.)

These are shocking statistics—the sort that should keep management awake at night. Following the PAM provided in this book might help you from becoming statistical roadkill. Using the standards of the International Organization for Standardization or ISO offers you guidance in this endeavor.

Now project management process capability management is not a panacea. Peter Drucker once said, “Nothing is less productive than to make more efficient what should not be done at all.” (From https://www.facebook.com/peterdruckerquotes). Our book will not tell you whether you are doing the right things, but it could tell you whether you are doing them the right way.

Introduction

This publication provides a process assessment model and related assessment guide based on ISO 21500:2012 *Guidance on Project Management* that is compliant with International Organization for Standardization (ISO)/International Electrotechnical Commission (IEC) 33000 series on process assessment.

It provides the foundation for the assessment of an organization's project management processes against international standards. Using this evidence-based approach, organizations may solicit improvements in their project management processes or discover their present capability levels.

Why Is This Important?

While project management occurs in virtually all organizations in some form or other, far too many organizations lack a reliable, robust method for determining how well their staff follow project management processes. This book aims to supply such a methodology based on a standard process capability model.

What Is the State of Project Management Today?

So, what is a project and why manage it? According to ISO 21500: 2012, a project consists of a unique set of processes consisting of coordinated and controlled activities with start and end dates, performed to achieve project objectives. Achievement of the project objectives requires the provision of deliverables conforming to specific requirements. Every project has a definite start and end, and is usually divided into phases.

According to the Project Management Institute's *Pulse of the Profession 2017*, [update to 2017 Pulse] the number of high-performing organizations, those doing project management right, has dropped to 7 percent, a change from the 12 percent reported in 2012. High performing organizations are those that are utilizing proven project, program, and portfolio management practices that reduce risks, cut costs, and improve success rates of projects and program.[2]

According to a PwC report,[3] poor estimates, missed deadlines, scope changes, and insufficient resources comprise 50 percent of the reasons for project failure.

Another major challenge is ensuring the consistent application of defined processes, seen as a difficulty by some 38 percent of organizations.[4]

The Ontario government, among others, has had its fair share of failures—from the $4.5 million lost in the Court Information Management System project to the Electronic Health Record project purported to have cost around a billion dollars with little to show. Even Ronald McDonald was affected. Their Innovate Project, a real-time enterprise project, failed before it even got off the ground. In 2002, McDonalds wrote off $170 million.[5]

While there is a clear correlation between organizations with successful projects and the use of certified individuals from PRINCE2 (PRojects IN Controlled Environments) or PMI's Project Management

[2] http://www.pmi.org/-/media/pmi/documents/public/pdf/learning/thought-leadership/pulse/pulse-of-the-profession-2015.pdf

[3] https://www.pwc.com/us/en/people-management/assets/programme_project_management_survey.pdf

[4] http://www.pmsolutions.com/reports/State_of_the_PMO_2016_Research_Report.pdf

[5] http://www.kellogg.northwestern.edu/student/courses/tech914/summer2004/projectfailures/projectfailures/mcd.htm

Professional (PMP), the authors believe an additional element of success will include the practice of capability determination using the ISO 33000 series. Using this international standard as an assessment approach offers a consistent, world-wide method for ensuring the numerous elements of your business. Governance using COBIT™, ITIL, and now Project Management follow the same capability assessment program.

The approach used in this book is structured to assist organizations in improving their project management processes by providing either a clear understanding of their current capability levels or a clearly defined governance process for defining and attaining a desired capability level.

Why Change?

Successful organizations need little encouragement to further their success. Improvement is likely a part of their modus operandi. However, far too many organizations do not understand the extent of their project management capabilities and may rely on outdated or inconsistent methods for attempting improvement.

Using an internationally defined and approved standard for assessing capability can lead these organizations to a consistent level of performance, which typically results in lower project costs and more efficient operations. ISO standards are based on metrics, and applying these metrics to project management can assist in a better understanding of strengths and weaknesses, resulting in improvements for those that choose to do so.

While most organizations use a project management methodology, many are developed in-house rather than following a formal approach such as PMI's PMBOK or the Cabinet Office's PRINCE2. Regardless, determining how well these methodologies are followed can be daunting. This is where this publication can help. Your organization could determine its capability by following a well-designed International capability assessment approach.

Using ISO 33000 series standards offers a clear and concise methodology for capability determination. Why do we believe capability is the right way to go versus the Capability Maturity Model Integration (CMMI) maturity model? Firstly, our method is ISO-based, which follows the rigor required to produce a world-wide approved standard.

Secondly, process capability focuses on the performance within each process, defining whether it satisfies its performance and quality objectives, while also producing output that is within the desired specifications.

Following this approach, organizations might achieve better results within their projects as project managers, and PMOs might begin to help ensure successful implementation of project management processes.

What Is the Purpose of This Book?

This book will provide a clear, concise, and repeatable methodology for ensuring improvement within each project process and will define current project management capability levels.

It is intended to serve as a guide to both an understanding of capability terminology and the steps involved in pursuing capability. It also functions as a detailed step-by-step approach to performing capability assessments of your project management processes. As Martin Luther King, Jr., said, "You don't have to see the whole staircase, just take the first step."

What Are the Potential Benefits?

See Table 1 for a review of several benefits to using a well-structured methodology such as ISO 33000.

Table 1 Benefits of ISO/IEC 33000

Consistency	Using ISO 33000 series helps define a clear path for improving project management processes. This provides for predictable project processes that are implemented to a defined level of capability.
Replicability	Project management processes, based on these ISO standards, offer organizations an easily replicated formula for success based on a keen understanding of how each process is performing as projects are managed.
Improvement	Using this methodology will help organizations improve their overall project management system, enabling better end results and lowering the risk of project failures.
Repeatability	Using the ISO/IEC 33000 series provides organizations with repeatable steps ensuring all project management processes function in a similar way at a specific capability level according to the organization's needs.

Does This Method Conform to ISO 15504?

ISO 33000 series replaces the older but still functional ISO 15504 series upon which the CRP Henri Tudor research center based its ITIL process assessment book. Others followed a similar path. Since the ISO 15504 series was replaced around 2015, this book chose to conform to the newer ISO/IEC 33000 series. This newer version follows a slightly different structure with several sections of the older version being combined within the new series. For example, in the new series, several of the standards combine parts of ISO/IEC 15504-2 and ISO/IEC 15504-7.

ISO/IEC 33001 provides an excellent table describing the relationships between the new and older series.

PART I
Process Reference Model

So, as Lao Tzu once said: "A journey of a thousand miles starts with a single step." Fortunately, you don't have that far to go. If you are knowledgeable of project management best practices, then much of Part I is very familiar to you.

1

The Standards

The first step is to understand the various organizations and standards we draw upon in this book. There is no time like the present, so let's get started.

What Is the International Organization for Standardization?

You may or may not have intimate knowledge of the International Organization for Standardization or ISO.[1] The ISO is an international organization focused on developing and promoting voluntary standards across a wide and expanding range of disciplines. Started in 1947, the ISO has published over 21,000 standards offering product and service specifications, as well as codifying "best practice," all with the aim to make your organization and others more efficient, effective, and cost-effective. In 1951, ISO issued the first standard, or recommendation as the ISO called them then, entitled ISO/R 1:1951 *Standard reference temperature for industrial length measurements.*

Around 163 countries have national standards bodies that support the ISO as either full or corresponding members. Should you have an interest, ISO publishes a list of member bodies at https://www.iso.org/members.html. For example, you will see that the Standards Council of Canada (SCC) represents Canada, the American National Standards Institute (ANSI) represents the US, and the British Standards Institution (BSI) represents the UK.

[1] You may wonder why it is not IOS. Well, ISO headquarters is in Geneva, CH where they speak French primarily. According to ISO.org, "Because 'International Organization for Standardization' would have different acronyms in different languages (IOS in English, OIN in French for Organisation internationale de normalisation), our founders decided to give it the short form ISO. ISO is derived from the Greek isos, meaning equal." Also, as you may know, when you put ISO in front of something; such as isometric, isobar, isotherm, it means "equal." This is apropos as all countries are equal in the eyes of the ISO. But as we used to say at a "Big Few" firm, there are partners, then there are partners. Oh, and CISCO and Apple are sitting on IOS trademarks or variants.

The ISO strives to end "technical nationalism" and democratize technical information through the provision of technical standards. ISO 21500 is an excellent example.[2]

Our book is comprised of two major parts. Part I provides the definition of the Process Reference Model based on ISO 21500, while Part II provides an explanation of the process capability assessment. ISO 33063, which provides the process capability assessment, is introduced in Part II. Together, the process reference model (Part I) and the assessor guide act as state-of-the-art process improvement.

Then What Is ISO 21500?

The ISO started working on ISO 21500:2012, *Guidance on Project Management*, in 2007 and released it in 2012. As the title suggests, ISO intended to provide generic guidance, explain core principles and what constitutes good practice in project management. The guidance standard was not meant for the purposes of certification. Perhaps ISO will change it to a project management system standard in a future release and promote certification.

The project management technical committee was held by the *American National Standards Institute* or ANSI.[3] ANSI had previously approved four standards based on Project Management Institute's (PMI) bodies of knowledge. Of interest to us is ANSI/PMI 99-001-2008, *A Guide to the Project Management Body of Knowledge—3rd Edition (PMBOK®Guide—3rd Edition)*, which was a revision and re-designation of ANSI/PMI 99-001-2004: 11/20/2008. ANSI put forward its standard for consideration. You will see a lot of similarity between the *PMBOK* third and fourth editions and ISO 21500:2012. While the authors were writing this book, ISO 21500 was under revision; neither revision should affect what we offer here. Our hope is that the updates will supplement the process detail.

---

[2] Though one should wonder about this ISO goal. The PMBOK 6th Edition costs $67.04 USD (your price may vary) on Amazon.com and is 756 pages long, which is about 9 cents per page. ISO 21500:2012 costs $158 CHF (or about $162.94 USD) and is 36 pages long, which is a whopping 453 cents per page.

[3] If you are an American, you would purchase your standards from ANSI when available. If they are not available, then you would purchase them directly from the ISO.

ISO plans that ISO 21500 be the first in a family of project management standards. There is the opportunity, for instance, to provide guidance on earned value, project complexity, and project risk management.

ISO 21500 aligns with other related standards such as ISO 10005:2005, *Quality management systems—Guidelines for quality plans*, ISO 10006:2017, *Quality management systems - Guidelines for quality management in projects*, ISO 10007:2017, *Quality management systems - Guidelines for configuration management*, and ISO 31000:2018, *Risk management—Principles and guidelines.*

What Is the Value of ISO 21500?

While management is asking for more agility, your organization is facing a competitive and increasingly hostile, complex environment. It is tough, but research shows that projects managed using structured processes, leveraging "best practices," consistently show higher performance than those that do not in the following areas:

- Up-front planning—whether using waterfall, spiral, or agile methodologies—helps projects deliver value.
- Using project management "best practices" helps organizations deliver quicker.
- Using proactive project management processes, such as those in ISO 21500 and PMBOK, usually results in less surprises during project execution.
- Delivering a quality project on time within budget leads to improved customer satisfaction and less rework, which is often a type of waste in organizations.

You could use ISO 21500 for the following reasons:

- Use as a reference in an audit, a review or an assessment. We are using it here for the purposes of an assessment as defined in ISO 33000 standards, but you could perform an internal or external audit to show "compliance" with the standard. Obviously, you could perform a review of project management using ISO 21500, which usually has less prescribed processes than an audit or assessment.

- To promote communication and trust. Each project has different parties with different backgrounds and experience, including the project sponsor, project director, project manager, project team, line of business, customers, and users. These constituencies all have their own jargon. Sometimes when they try to communicate among themselves, it seems like some are attempting to speak a foreign language. Work breakdown schedule, what is that? Everyone within your organization could use the terminology and concepts from ISO 21500 to promote a common understanding and improve communication; thereby, leading to greater cooperation and trust. It is equally important to have a common language for projects when you have multilingual, multinational, and multidisciplinary projects and multiple project methodologies. ISO 21500 could be the glue to bind it all together.
- To provide a linkage between various guidance. ISO 21500 provides an anchor and commonality between various guidance such as PMBOK, PRINCE2, and IPMA Competence Baseline (ICB).
- Use to link project processes and business processes. Most ISO standards supplement ISO 9001, the quality management system for an organization, and ISO 21500. The assessment methods introduced in this book are no exception as the standards encourage process improvement.
- Use as a checklist for project team members. When initiating, planning, executing, and closing projects, project teams and especially the project manager, could use ISO 21500 to ensure that inputs are used, activities are performed, and outputs are delivered.

What Is the Difference between a Standard and Guideline?

If you are not familiar with ISO terminology, you might not understand the difference between the various types of ISO standards. Logically, you would think that any ISO standard is on the same footing or of the same importance. As we said, ISO standards are voluntary. However, an industry could adopt the standard, making it mandatory for the sector (should it have that power). Additionally, a government could do the same (and they generally have the power to do so, to wit U.S. Sarbanes–Oxley Act of 2002).

In the ISO world, there are:

1. Descriptive or informative standards; and,
2. Prescriptive or normative standards.

Generally, people tend to think of a standard as prescriptive, that is, those shalt do or shalt not do something statements. Those statements are prescriptive. ISO guides and guidance, however, fall into the descriptive. Therefore, ISO 21500 is descriptive or informative in nature.

Why Use ISO 21500?

Without a structured method for project management, it is difficult for an organization to excel at delivering a project on time, on budget, and meeting quality requirements. In fact, you might see any of the following signs listed below.

- Projects managers "wing it" because of inadequate standard methods for project management processes and techniques.
- Project management is regarded as a cost center with little or no value.
- Project work is undertaken without careful or appropriate planning.
- Project resources are seconded to projects while still accomplishing their normal, full-time duties.
- Project budgets do not include internal resource costs and are considered a "sunk cost."
- Project management office does not exist or does not have insight into the true benefits of all current projects.
- Proactive project management costs and activities are not included in project plans.
- Projects meet targets through herculean efforts causing project team member stress and burnout.

Because the World Bank estimates that at least one-fifth of the world's GDP or roughly $12 trillion USD[4] will be spent on project-based work, organizations must become more effective at managing projects. ISO 21500 and this book aim to help you achieve that goal.

Are There Other Standards and Methods for Project Management?

Absolutely, there are many project management standards and methodologies. This was one of the issues ISO 21500 meant to address by creating an international standard that member bodies accredit. At its core, not only did the standard need accreditation by member bodies, its inherent relevancy allowed for widespread project management community support and adoption.

When developing ISO 21500, the following were either offered as models or were used as reference material:

- *A Guide to the Project Management Body of Knowledge* (PMBOK Guide), Third Edition, Project Management Institute Inc., U.S. fix to fifth edition.
- BS 6079 and BS ISO 15188:2001 *Project Management.*
- DIN 69901 *Project Management: Project Management Systems*, Germany, 2007.
- ICB Version 3.0 (*IPMA Competence Baseline*), International Project Management Association.
- ISO 10006 *Quality Management Systems—Guidelines for Quality Management in Projects*, ISO.
- PRINCE2, Cabinet Office, U.K.

The closest guidance to ISO 21500 is PMI's PMBOK. The processes in ISO 21500 are almost identical to the processes in earlier PMBOK Guides, the set of 10 subjects in ISO 21500 follows the set of Knowledge areas in the PMBOK Guide. There are 39 Processes in ISO 21500 and 47processes in the PMBOK Guide. Thirty-three processes in ISO 21500 are almost the same as the processes in PMBOK

[4] Obviously, the dollar figure could fluctuate and grow as the world economy grows, but the one-fifth number is probably a good rule of thumb. Think of your organization and all the projects under way at any given point in time.

Guide. The main difference between ISO 21500 and the PMBOK Guide is that ISO 21500 does not provide tools and techniques. ISO 21500 provides high-level description of concepts and processes and can be used by any type of organization, including public, private or community organizations, and for any type of project, regardless of complexity, size and duration. This is where the guides in this book provide the reader value. We have taken the activities and artefacts in the PMBOK and mapped them to ISO 21500 so that you could perform a process capability assessment.

There also are the Association for Project Management (APM) qualifications. The APM is a registered charity with over 22,000 individual and 550 corporate members, making it the largest professional body in the United Kingdom. APM aims to develop and promote the professional disciplines of project management and program management, through a program called the "FIVE Dimensions of Professionalism." The APM Introductory Certificate (IC) assesses fundamental knowledge in project management and about the profession of project management.

While the ISO 21500 standard, the APM certifications and the PMBOK tell you what you ought to know as a project manager, PRINCE2 tells you what you ought to do. It is a methodology that maps very closely to the guidance mentioned above. You will see PRINCE2 referenced in process details.

At the end of the day, ISO 21500 is a common reference that bridges different methods, practices, and models by providing a common language for project management. It offers a single, global standard and an overarching document for project management.

What Is the Structure of ISO 21500?

The ISO 21500 standard follows this structure:

- Clause 1 Scope. The scope of ISO 21500 is simply the management of projects in "most organizations most of the time."
- Clause 2 Terms and definitions. There are 16 project management terms with definitions. There are many terms that could have been included, but the developers only included those terms not properly defined in other ISO standards.

- Clause 3 Project management concepts. Clause 3 describes concepts that are important in the execution of most projects. They are:

    - 3.1 General
    - 3.2 Project
    - 3.3 Project management
    - 3.4 Organizational strategy and projects
    - 3.5 Project environment
    - 3.6 Project governance
    - 3.7 Projects and operations
    - 3.8 Stakeholders and project organization
    - 3.9 Competencies of project personnel
    - 3.10 Project life cycle
    - 3.11 Project constraints
    - 3.12 Relationship between project management concepts and processes

    These concepts and their relationships are well known to project managers.
- Clause 4 Project management processes. Clause 4 identifies the recommended project management processes for a project or project phase.
    - 4.1 Project management process application.
    - 4.2 Process groups and subject groups. The project team would normally repeat process groups in each project phase.
    - 4.3 Processes. This is the majority of the standard and suggests the processes for the process reference model and the process capability assessment.
- Annex A (Informative) Process group processes mapped to subject groups.

For our purposes, we are interested in the key project processes.

Since the standard is meant for various industries and sectors, the processes are generic and useful for any project in any organization of any size. The project manager and stakeholders must decide at the beginning of a project or project phase what processes in Section 4.3 of ISO 21500 provide greatest return to the organization and tailor that method to their environment and project. You need not apply the processes uniformly on all projects or all project phases; use only those that align with your organization's goals. You will see later that you must align the process with the business goals when we look at the generic work practices. Part of this tailoring process involves the project manager and stakeholders determining the rigor that needs to be applied for each process.

What Is Process Capability?

Many people confuse the concepts of process maturity and process capability.

Process maturity means that whatever an organization is doing, it does in a well-documented way, and everyone knows what is expected of them and performs accordingly. Using mature processes, performance is not dependent on heroes and decisions are made on proper situational analysis. To comment on organizational maturity, one must first determine process capability.

A process is capable when it satisfies its specified product quality, service quality, and process performance objectives. A capable process consistently produces output that is within specifications. These specifications come from the client, industry benchmarks or service-level agreements. Execution of a capable process always gives predictable results. It is not free of defects, but the defects are known or are within an acceptable level. Most of you have heard of Six Sigma, which translates to 3.4 defects per million opportunities (DPMO). We will see that this is one of the standard's capability level and some quality tools are referenced. With the new ISO 33000 standards, the ISO is trying to align capability assessments with standards such as ISO 13053-1:2011, *Quantitative methods in process improvement—Six Sigma—Part 1: DMAIC methodology,* and ISO 13053-2:2011, *Quantitative methods in process improvement—Six Sigma—Part 2: Tools and techniques.*

We define process capability models in a reference scheme; that is, a process reference model, the processes, the pre- and post-conditions of the application, and the resulting work products. An associated process assessment model defines a hierarchical evaluation scheme to assess the maturity and capability on certain levels. You will learn about the assessment model in Part II of this book.

Process capability levels apply to an organization's process improvement achievement in individual process areas. These levels are a means for incrementally improving the capability of processes corresponding to a given process area. At each level, there are process indicators, or what many people would call "best practices"[5] and inputs and outputs. They classify the performance of project management processes of a certain process area done by an organization, organizational department, or project. As Carly Fiorina, former HP Chair, once said, "The goal is to turn data into information, and information into insight."[6]

Since project management is a collection of processes, it is important to make sure the processes are capable. We produce the best product or service when we start with the best inputs and use a capable process to transform the inputs.

But How Do We Determine Capability?

There are many ways, such as using Capability Maturity Model Integration (CMMI)[7] and Six Sigma to name two. While the former is useful for assessing software development maturity and the latter for assessing process stability and capability and solving process problems, they are not necessarily the best tools to assess project management process capability. The ISO 33000 standard family; and the former ISO 15504 series, is the answer.

Two parts of process capability are: (1) measure the variability of the output of a process, and (2) compare that variability with a

---

[5] Generally, we use legitimate best practices, which are usually industry-related, for processes where things are known, we have lots of data, and the cause-and-effect relationships are repeatable, perceivable, and forecastable. In turn, we use good practices where things are knowable, we have less data, and the cause-and-effect relationships are separated over space and time.

[6] Source: http://www.hp.com/hpinfo/execteam/speeches/fiorina/04openworld.html

[7] The CMMI Institute is now a subsidiary of ISACA.

proposed specification or product tolerance. When doing process capability analysis, you must compare the performance of a process against its specifications. With a manufacturing process, this is a relatively easy task as we have many instances or examples of the process inputs and outputs. That is why we have well-established tools and techniques such as Six Sigma. For business processes, this is not so easy. In business, this is not atypical: we need to find examplars from the physical world and use them. This is not as difficult as it may sound. The specifications in ISO 21500 are the inputs, best practices and outputs, what we call artefacts. We say that a process is capable when most of the possible variable values are used or created.

NOTE

Process instance is a term used to describe a set of process activities or certain types of processes. It is the items that make up a process. In the Establish (ES) domain, process 2—Set policies, processes, and methodologies, an instance might be one policy or a single methodology.

The concept of process capability determination by using a process assessment model is based on a two-dimensional framework. The first dimension is provided by processes defined in a process reference model (process dimension). We will use ISO 21500 as the sole source to develop our process reference model (PRM).

The second dimension consists of capability levels that are further subdivided into process attributes (capability dimension). The process attributes provide the measurable characteristics of process capability. We will use ISO/IEC 33020 to develop our capability dimension.

Process capability indicators are the means of achieving the capabilities addressed by the considered process attributes. Evidence of process capability indicators supports the judgment of the degree of achievement of the process attribute. The capability dimension of the process assessment model consists of six capability levels matching the capability levels defined in ISO/IEC 33020.

Following ISO/IEC 33020, we describe the process capability indicators for the nine process attributes included in the capability dimension for process capability levels 1–5. Each of the process attributes in our process assessment model is identical to the process attribute defined in the process measurement framework. The generic practices address the characteristics from each process attribute. The generic resources relate to the process attribute as a whole. Process capability level 0 does not include any type of indicators, as it reflects a non-implemented process or a process that fails to partially achieve any of its outcomes. It may seem complicated, but we will walk you through the process reference model and the assessment model before we finish.

Yes, there is some subjectivity in our assessment. No matter how objective your measures, humans introduce some subjectivity. Even with something as objective as a Sigma level, a human still needs to assess the results. Is the level good enough or not?

But You Can't Get Certified on ISO 21500, Correct?

Since ISO 21500:2012 is a guidance document, it is not intended to be used for certification or registration purposes. You cannot get certified ISO 21500 as an organization. However, there are several organizations, such as the Professional Evaluation and Certification Board and ProjectManagers.org, certifying individuals. The former offers the *Certified ISO 21500 Lead Project Manager* (https://pecb.com/iso-21500-lead-project-manager-certification) and *Certified ISO 21500 Lead Auditor* (https://pecb.com/iso-21500-lead-auditor-certification) certifications. Similarly, the latter offers *Certified Project Manager in ISO 21500:2012* (http://iso21500.projectmanagers.org/p/cert-holders.html) certifications. Theoretically, any organization could offer such certifications.[8]

Although organizations cannot become ISO 21500–certified, the capability of the processes could still be *assessed*, which is much

[8] Should you have an interest in such a certification, make sure the training organization itself is certified ISO/IEC 17024. This certification means that the organization is an accredited Personnel Certification Body under ISO/IEC 17024—*Requirements for bodies operating certification of persons.*

different than performing an audit for certification. Terms such as audit, review, and assessment have legal meanings.

The assessment and self-assessment described herein are not certifications, but rather a formal or self-declaration of how well your organization applies the principles and guidelines of ISO 21500. In the end, you should use the results of the assessment or self-assessment to become better at managing projects. You should strive for progress, not perfection, in your project management processes. To quote Robert Collier, American author, "Success is the sum of small efforts, repeated day-in and day-out."[9]

What Are the ISO 33000 Standards?

You use the collective ISO 33000 family to assess process quality characteristics, such as process safety, efficiency, effectiveness, security, integrity and sustainability.

The ISO 33000 family is architected as shown in Table 1.1.

The standards of particular interest to us are:

- ISO/IEC 33001:2015 *Information technology: Process assessment—Concepts and terminology*
- ISO/IEC 33002:2015 *Information technology: Process assessment—Requirements for performing process assessment*
- ISO/IEC 33003:2015 *Information technology: Process assessment—Requirements for process measurement frameworks*
- ISO/IEC 33004:2015 *Information technology: Process assessment—Requirements for process reference, process assessment and maturity models*
- ISO/IEC 33014:2015 *Information technology: Process assessment—Guide for process improvement*
- ISO/IEC 33020:2015 *Information technology: Process assessment—Process measurement framework for assessment of process capability*
- ISO/IEC 33063:2015 *Information technology: Process assessment—Process assessment model for software testing*

[9] Source: http://www.quotery.com/quotes/success-is-the-sum-of-small-efforts-repeated-day-in/

Table 1.1 ISO 33000 Standards

THEME	STANDARD FOCUS		
OVERALL	*CONCEPTS AND TERMINOLOGY* (ISO/IEC 33001)		
REQUIREMENTS	ASSESSMENT PROCESS	MEASUREMENT FRAMEWORK	PROCESS MODELS
Principles	*Requirements for performing process assessment* (ISO/IEC 33002)	*Requirements for process measurement frameworks* (ISO/IEC 33003)	*Requirements for process reference, process assessment, and maturity models* (ISO/IEC 33004)
Element standards	Documented assessment processes (ISO 33030-33039); for example, *An examplar documented assessment process (ISO/IEC TS 33030)*	Frameworks for measuring process quality characteristics (ISO 33020-33029); for example, *Process measurement framework for assessment of process capability (ISO/IEC 33020)*	Process reference models (ISO 33040-33059); for example, *Safety extension* (ISO 33040) Process assessment models (ISO 33060-33079); for example, *Process assessment model for software testing* (ISO/IEC 33063) Maturity models (ISO 33080-33099); for example, *An Integrated Organizational Maturity Model for Software and Systems Engineering* (ISO 33081)
Application guides and supplements	*Guide on performing assessments* (ISO 33010) *Guide on defining a documented assessment process for assessment* (ISO 33011) *Process Assessment Body of Knowledge* (ISO 33016)	*Guide for process improvement* (ISO 33012) *Guide for process capability determination* (ISO 33013)	*Guide for constructing process reference models, process assessment models, and organizational maturity models for assessments* (ISO 33014)

Do not concern yourself that these standards fall under Information Technology. All organizations regardless of their industry are information based. You could not carry out a project without information. Manufacturers use specifications, which are information, to build products. Service organizations need to understand their customer's needs and meet complex billing requirements therefore relying on a Customer Information System.

This process assessment method does not allow making an assessment of organizational maturity. Moreover, no benchmarking data is available so far. ISO/IEC 33003 defines the requirements for a measurement framework. The aim of the standard is to improve one's comprehension of process measurement frameworks for process quality characteristics. Also, ISO/IEC 33003 talks about reflective and formative measurement models. The measurement model offered in this book is a formative measurement model, in that the measures:

- Are defining characteristics;
- Are not necessarily interchangeable;
- Do not have similar content;
- Do not necessarily co-vary with one another;
- Combine to form the model; and,
- Do not change should the model change.

We can see that ISO/IEC 33020 is the source for our process measurement model.

What Is the Structure of ISO/IEC 33020?

ISO 33020 follows this structure:

- Clause 1 Scope. Assessment of process capability in accordance with ISO/IEC 33002.
- Clause 2 Normative resources. Documents referenced in the standard.
- Clause 3 Terms and definitions. There are only two assessment terms with definitions. There are many terms that could have been included, but the developers only included those terms not properly defined in other ISO standards.

- Clause 4 Overview. Declares that the principles in the standards apply to any domain, such as project management.
- Clause 5 A process measurement framework for process capability. This clause defines the process measurement framework for the assessment of process capability conforming to ISO/IEC 33003 and includes the following sub-clauses.
  - 5.1 Introduction
  - 5.2 Process capability levels and process attributes
    - 5.2.1 Process capability Level 0: Incomplete process
    - 5.2.2 Process capability Level 1: Performed process
      - 5.2.2.1 PA 1.1 Process performance process attribute
    - 5.2.3 Process capability Level 2: Managed process
      - 5.2.3.1 PA 2.1 Performance management process attribute
      - 5.2.3.2 PA 2.2 Work product management process attribute
    - 5.2.4 Process capability Level 3: Established process
      - 5.2.4.1 PA 3.1 Process definition process attribute
      - 5.2.4.2 PA 3.2 Process deployment process attribute
    - 5.2.5 Process capability Level 4: Predictable process
      - 5.2.5.1 PA 4.1 Quantitative analysis process attribute
      - 5.2.5.2 PA 4.2 Quantitative control process attribute
    - 5.2.6 Process capability Level 5: Innovating process
      - 5.2.6.1 PA 5.1 Process innovation process attribute
      - 5.2.6.2 PA 5.2 Process innovation implementation process attribute
  - 5.3 Process attribute rating scale
  - 5.4 Process attribute rating method
    - 5.4.1 Rating method R1
    - 5.4.2 Rating method R2
    - 5.4.3 Rating method R3
  - 5.5 Aggregation method
    - 5.5.1 One dimensional aggregation methods
      - 5.5.1.1 One dimensional aggregation using arithmetic mean
      - 5.5.1.2 One dimensional aggregation using median

  - 5.5.2 Two dimensional aggregation methods
    - 5.5.2.1 Two dimensional aggregation using arithmetic mean
    - 5.5.2.2 Two dimensional aggregation using heuristics
- 5.6 Process capability level model
- Annex A (Informative) Conformity of the process measurement framework
- Annex B (Informative) Example of a process performance model
- Bibliography

How Do We Measure Capability?

For our assessment in Part II, we need a measurement scale. ISO/IEC 33020 does not provide a measurement scale for process attributes. Rather you must look to ISO/IEC 33003 subsection 4.5 *Scoring process attributes* that requires you use a measurement scale. Your scale could be nominal-, ordinal-, interval- or ratio-based. The section implies that should you rate a process attribute using a ratio scale, you should also use a ratio scale for measuring process attributes. However, the section also implies that should you rate a process attribute using an ordinal scale, you could use an ordinal, interval or ratio scale for measuring process attributes.

For the purposes of our assessment, we have chosen an ordinal scale; that is, levels 0–5. While you might consider the scale nominal-based, in that the numbers name the level just as the jersey number names a player in a sport, it is indeed ordinal-based because the numbers refer to a progressive order from Level 0 to Level 5.

Aggregation of Ratings

Because you will be using a formative rating model, refer to the aggregation of the measures to create a composite index rather than a composite rating. This statement is not important to understand the process reference model or to perform assessments, but it is important to academia. Practically this means that when rating a process attribute, you may aggregate ratings of the associated process attributes across all processes, and for any given attribute, you may aggregate the ratings for all instances of a process. The former is labelled vertical aggregation and the latter is horizontal aggregation. We address aggregation in Part II.

Does ISO 15504 Still Apply?

Our book relies on the newest version of Process Capability, the ISO/IEC 33000 series. There are, however, similarities to this version and the prior version, the ISO/IEC 15504 series. Many of the diagrams and basic information are similar in either version. However, the newer version updates and redefines concepts, it changes several of the attributes and adds rating aggregation methods among other items and is now five books rather than the seven in ISO/IEC 15504. While the books are not necessary for performing a project management capability assessment, they will provide a more in-depth understanding, so it is recommended that you purchase them from the ISO store.

The first book, ISO/IEC 33001:2015, *Information technology: Process assessment—Concepts and Terminology*, provides the key terminology relating to process assessment. It gives you high-level information on the concepts of process assessment, as well as information on using process assessment for evaluating the achievement of process quality characteristics.

The second book in the series, ISO/IEC 33002, *Information technology—Process assessment—Requirements for performing process assessment* provides the reader with the concepts of assessment itself, including the requirements needed to achieve objective, repeatable and consistent results. It also gives you information on the three classes of assessment.

The third book, ISO/IEC 33003:2015, *Information technology: Process assessment—Requirements for process measurement frameworks* provides the requirements for your measurement framework using defined process attributes.

Book four, ISO/IEC 33004:2015, *Information technology: Process assessment—Requirements for process reference, process assessment and maturity models* gives the reader the requirements for process reference models, process assessment models, and maturity models. It shows you the relationship between the classes involved, the use of common sets of assessment indicators of process performance and process quality, and the integration of process reference models and process measurement frameworks used to establish process assessment models.

Book five, ISO/IEC 33020:2015, *Information technology: Process assessment—Process measurement framework for assessment of process capability* is the final one in the series and provides the requirements for measuring your process capability. The measurement framework enables the user to perform self-assessments, and produce a set of process attribute ratings, and hence derive a process capability level.

2
The Process Assessment Model

We have identified all the standards and their relationships; now it is time to use them.

What Is the ISO 21500 Process Assessment Model?

An important component of this ISO 21500 assessment program is the ISO 21500 Process Assessment Model (PAM), which is shown in Figure 2.1.

As you could see in the figure, the PAM is comprised of the process reference model (PRM), which defines Level 1 base requirements, and the measurement framework, which determines the capability levels. The PAM combines the process descriptions in ISO 21500, which we have supplemented with PMBOK and PRINCE2 process detail, and ISO 33020, which describes the assessment process.

Part II of this book will cover the assessment process and comprises the Self-Assessor and Assessor guides.

The ISO 21500 PAM, comparable to other assessment models, is a two-dimensional model as shown in Figure 2.2.

In the figure, you could see that one dimension is the process dimension. We decompose processes into 2 process areas, 7 domains, and 45 processes. The processes themselves are decomposed into activities or

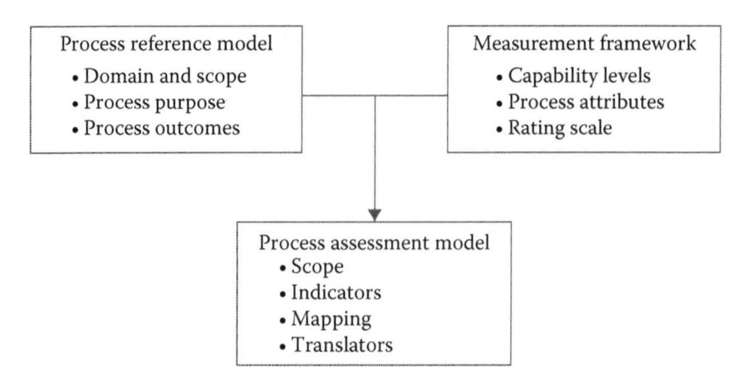

Figure 2.1 Process assessment components.

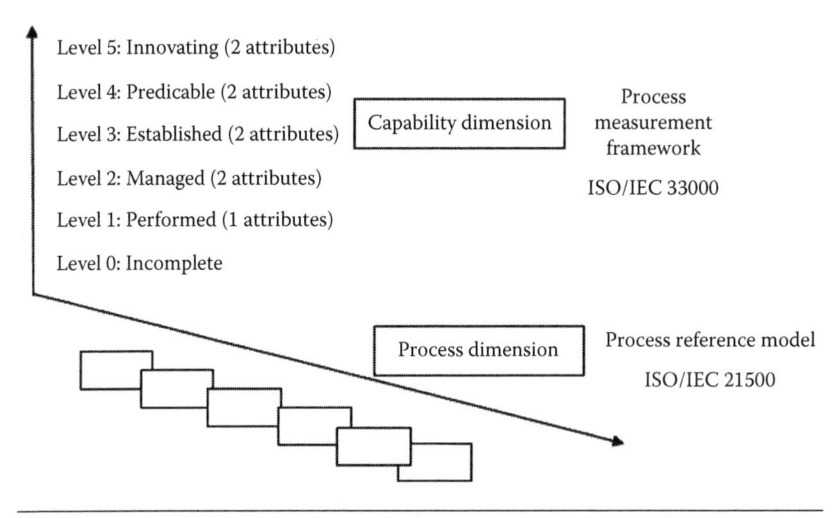

Figure 2.2 Process dimensions. (Permission to use extracts from ISO/IEC 33063:2015 was provided by the Standards Council of Canada [SCC]. No further reproduction is permitted without prior written approval from SCC.)

best practices. You also see the capability dimension, which is a set of nine process attributes distributed over the capability levels.

Our process dimension uses ISO 21500 as the basis for the PRM. Figure 2.3 shows the PRM. We added the governance domains to agree with Section 3.6 *Project Governance*.

Governance

Establish	ES01 Define the project management framework ES02 Set policies, processes, and methodologies ES03 Set limits of authority for decision-making	
Monitor	MO01 Ensure project benefits MO02 Ensure risk optimization MO03 Ensure resource optimization	

Management

Initiate	IN01 Develop project charter IN02 Identify stakeholders	IN03 Establish project team
Plan	PL01 Develop project plans PL02 Define scope PL03 Create work breakdown structure PL04 Define activities PL05 Estimate resources PL06 Define project organization PL07 Sequence activities PL08 Estimate activity durations	PL09 Develop schedule PL10 Estimate costs PL11 Develop budget PL12 Identify risks PL13 Assess risks PL14 Plan quality PL15 Plan procurements PL16 Plan communications
Implement	IM01 Direct project work IM02 Manage stakeholders IM03 Develop project team IM04 Treat risks	IM05 Perform quality assurance IM06 Select suppliers IM07 Distribute information
Control	CO01 Control project work CO02 Control changes CO03 Control scope CO04 Control resources CO05 Manage project team CO06 Control schedule	CO07 Control costs CO08 Control risks CO09 Perform quality control CO10 Administer procurements CO11 Manage communications
Close	CL01 Close project phase or project	CL02 Collect lessons learned

Figure 2.3 Process reference model.

Figure 2.3 shows the following two process areas:

1. Project governance involves setting direction and making decisions about and monitoring project constraints and benefits.
2. Project management (PM) involves delivering on the project direction within project constraints while delivering agreed-upon benefits.

Process areas decompose into process domains. Domains are a logical grouping of processes. The PRM has 45 processes that describe the PM life cycle.

Governance Domains

As per Section 3.6, page 6, "governance is the framework by which an organization is directed and controlled." The section offers guidance on

- Management structure.
- Policies, processes and methodologies.
- Authority limits.
- Stakeholder accountabilities and responsibilities.

Within the Project Governance area, there are two domains:

1. The Establish (ES) domain provides us with the governance processes that set the background and framework for PM in your organization.
2. The Monitor (MO) domain provides us with the governance processes that oversee the key inputs and outputs of the project.

Governance Processes

Within the ES domain, there are three processes:

1. Define the PM framework
2. Set policies, processes, and methodologies
3. Set limits of authority for decision-making

Within the MO domain, there are three processes:

1. Ensure project benefits
2. Ensure risk optimization
3. Ensure resource optimization

Management Domains

Within the Project Management area, there are five domains:

1. Initiate
2. Plan
3. Implement
4. Control
5. Close

Management Processes

The Initiate (IN) domain provides us with the management processes that initiate a project phase or project, including defined objectives and authorization for the project manager to start the project. Within the Initiate domain, there are three processes:

1. Develop project charter
2. Identify stakeholders
3. Establish project team

The Plan (PL) domain provides us with the management processes that develop baseline project detail to manage the project, monitor its performance, and ensure benefit delivery. Within the Plan domain, there are 16 processes:

1. Develop project plans
2. Define scope
3. Create work breakdown structure
4. Define activities
5. Estimate resources
6. Define project organization

7. Sequence activities
8. Estimate activity durations
9. Develop schedule
10. Estimate costs
11. Develop budget
12. Identify risks
13. Assess risks
14. Plan quality
15. Plan procurements
16. Plan communications

The Implement (IM) domain provides us with the management processes that deliver, according to the project plan, the benefits. Within the Implement domain, there are seven processes:

1. Direct project work
2. Manage stakeholders
3. Develop project team
4. Treat risks
5. Perform quality assurance
6. Select suppliers
7. Distribute information

The Control (CO) domain provides us with the management processes that monitor, measure, and compare project performance to the project plan and take corrective action or escalate issues as necessary. Within the Control domain, there are 11 processes:

1. Control project work
2. Control changes
3. Control scope
4. Control resources
5. Manage project team
6. Control schedule
7. Control costs
8. Control risks

9. Perform quality control
10. Administer procurements
11. Manage communications

The Close (CL) domain provides us with the management processes that officially terminate the project and document lessons learned. Within the Close domain, there are two processes:

1. Close project phase or project
2. Collect lessons learned

Process Groups

The authors of ISO 21500 refer to domains as process groups. Similar to domains defined in ISO 33001, process groups consist of processes that are applicable to any project phase or project. They are a logical grouping of processes. Grouping processes within a domain may help you in using the reference model or performing an assessment; for example, you could talk about the planning processes vis a vis the initiating processes, or you could assess all planning processes. But your assessment is not area or domain focused: it is process focused.

While not specifically defined in ISO 21500, it is not difficult to abstract process areas from the current standard. PRINCE2 provides good guidance on the areas, but additional information is provided in ISO 21500 and the PMBOK.

Subject Groups

In addition, the guideline (and PMBOK) categorizes them into the following subject groups:

1. Integration: The processes that cut across the project to tie it together and identify, define, combine, unify, coordinate, control and close project activities and processes.

2. Stakeholder: The processes that identify and manage the various interested parties, such as the business, user and supplier.
3. Scope: The processes that identify and define the required work and deliverables.
4. Resource: The processes that identify and acquire sufficient project resources, such as people, facilities, equipment, tools and funds.
5. Time: The processes that schedule and monitor project activities and progress.
6. Cost: The processes that develop the budget and monitor expenditures to control project costs.
7. Risk: The processes that identify and manage threats and opportunities and their impact on the project objectives.
8. Quality: The processes that establish quality requirements and assure they are met.
9. Procurement: The processes that acquire the external necessary resources—products, services, equipment, surge staff, et cetera—and manage supplier relationships.
10. Communication: The processes that determine, manage and distribute the project information required for every stakeholder.

While the subject groups are informative, they do not add anything to our taxonomy or the ultimate process capability assessment and are thus only mentioned here. When performing a process capability assessment, one might find it useful when writing the final report to discuss integration planning processes and resource planning processes. Table 2.1 shows the project management processes (but not the project governance processes) from ISO 21500 by subject and process groups.

Section 3 in Part I, provides the results of a best practices survey to build the project management process reference model. We leveraged selected practices from process capability models (such as SW-CMM,

Table 2.1 Processes Mapped to Subject Groups and Process Groups

SUBJECT GROUPS	PROCESS GROUPS				
	INITIATING	PLANNING	IMPLEMENTING	CONTROLLING	CLOSING
Integration	4.3.2 Develop project charter	4.3.3 Develop project plans	4.3.4 Direct project work	4.3.5 Control project work 4.3.6 Control changes	4.3.7 Close project phase or project 4.3.8 Collect lessons learned
Stakeholder	4.3.9 Identify stakeholders		4.3.10 Manage stakeholders		
Scope		4.3.11 Define scope 4.3.12 Create work breakdown structure 4.3.13 Define activities		4.3.14 Control scope	
Resource	4.3.15 Establish project team	4.3.16 Estimate resources 4.3.17 Define project organization	4.3.18 Develop project team	4.3.19 Control resources 4.3.20 Manage project team	
Time		4.3.21 Sequence activities 4.3.22 Estimate activity durations 4.3.23 Develop schedule		4.3.24 Control schedule	
Cost		4.3.25 Estimate costs 4.3.26 Develop budget		4.3.27 Control costs	
Risk		4.3.28 Identify risks 4.3.29 Assess risks	4.3.30 Treat risks	4.3.31 Control risks	
Quality		4.3.32 Plan quality	4.3.33 Perform quality assurance	4.3.34 Perform quality control	
Procurement		4.3.35 Plan procurements	4.3.36 Select suppliers	4.3.37 Administer procurements	
Communication		4.3.38 Plan communications	4.3.39 Distribute information	4.3.37 Manage communications	

ISO/IEC 15504-5, CMMI, CMMI-DEV, OPM3, COBIT, and eSCM-SP/CL), other reference models (such as ISO 9001, PMBOK, ISO/IEC 12207, and SWEBOK), project management methodologies (such as PRINCE2 and Agile DSDM), and other practice areas (such as RUP, DevOps and Lean) to further enhance the ISO 21500 information.

3

The Process Dimension

This section defines the process dimension and all the associated project management (PM) processes and the process performance indicators, which is also called the process capability dimension. These two dimensions form the overall picture of process assessment based on ISO/IEC 33000 (see Figure 2.2).

The Process Dimension

This dimension deals with the processes that you may wish to assess.

In our earlier diagram, the Process Dimension forms one of the two dimensions involved. In Figure 3.1, you can see the Process dimension and the key project management domains defined according to ISO/IEC 21500.

As we stated earlier, that means the processes related to ISO 21500:2012, *Guidance on Project Management*. How will this help? As you can see in the use case below, assessing your PM processes can lead to Improvement or current Capability determination (Figure 3.2).

Undergoing an assessment of your project processes affords you the opportunity to ensure they are being performed consistently, as determined by your organization against the capability levels described in this book.

You will first want to decide why you are assessing the processes. Is it to achieve better performance? Or, is to determine the current capability level of your PM processes and to determine a level of achievement that fits your organizational needs? Either way, a capability assessment will help ensure that your PM processes are fully understood and are operating at the level of competence required.

NOTE

If you choose to use a different project management model or methodology than ISO/IEC 21500, such as PRINCE or PMBOK, then you need to first map the preferred methodology against the ISO/IEC 21500 processes. Then, you will need to support any that do not match with process purposes, outcomes, base practices and the work products associated with those processes.

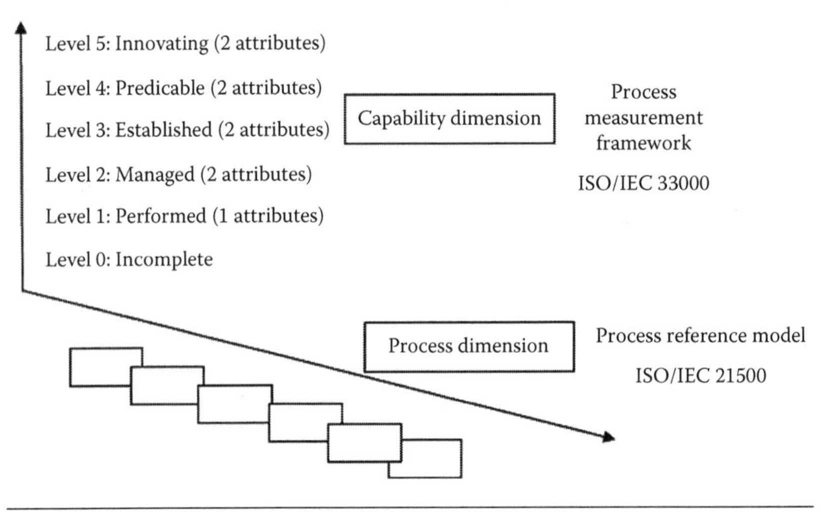

Figure 3.1 PM process dimension.

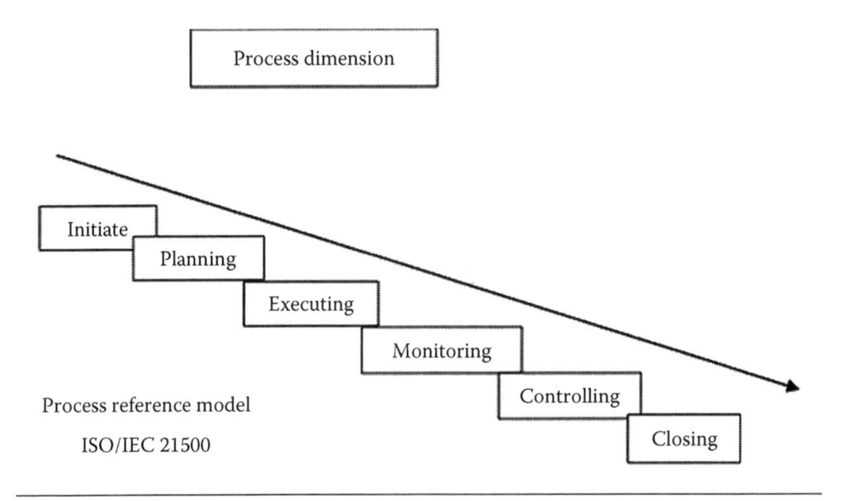

Figure 3.2 Process assessment use case.

You will find the individual processes are all described the same way: the same way you will find it described in any process reference model. There is a process name, description and purpose; outcomes; governance or management base practices; inputs and outputs; and references. In Section 4.1 of ISO 21500, the inputs and outputs are primary inputs and outputs, meaning you may decide to augment your assessment with either the PMBOK 6th Edition or PRINCE2 2017 and your expertise.

The base or generic practices and work products constitute the set of individual process performance indicators. Achieve these indicators and gain the capability level. In Part II, you will see how to use the indicators to perform the process capability assessment and support your ratings. The indicators are not exclusive, rather they are the minimum set needed to meet the process purpose. Some artifacts; such as the project plan, are shown at various levels and require more and more specificity as you move up the levels.

While you will complete your assessment based on the processes that support organizational goals and objectives, any initial assessment should focus on the core planning processes. "If you fail to plan, you plan to fail" is entirely appropriate in project management—you should spend 80 percent of your time in planning and 20 percent in execution.

Inputs and Outputs

There are several output work products in the various processes that are used as inputs to all processes. Rather than repeat them as an input many times, see the information provided in Table 3.1.

Where you see any of output work products specifically mentioned as an input to a governance or management process, it is done so because the input is a primary input and necessary to meet the process purpose.

The next section provides detail about the process reference model.

Table 3.1 Inputs to More than One Process

OUTPUTS TO ALL PROCESSES	
OUTPUT WP	OUTPUT DESCRIPTION
ES01-WP1	Project management guidelines
ES01-WP3	Definition of organizational structure and functions
ES02-WP1	Project management policy
MO02-WP1	Risk management policy
IN02-WP1	Stakeholder register
CL02-WP1	Lessons learned document
OUTPUTS TO ALL GOVERNANCE PROCESSES	
OUTPUT WP	OUTPUT DESCRIPTION
ES03-WP1	Authority limits
MO01-WP2	Benefits review plan
IN01-WP1	Project charter
OUTPUTS TO ALL MANAGEMENT PROCESSES	
OUTPUT WP	OUTPUT DESCRIPTION
ES02-WP2	Project management methodology
ES02-WP3	Project management process definition documents
ES02-WP4	Project documents
PL01-WP1	Project plans
PL02-WP1	Scope statement
PL03-WP1	Work breakdown structure
PL06-WP2	Project organization chart
PL12-WP1	Risk register
PL14-WP3	Quality management standards
PL16-WP1	Communications plan

Establish Domain

1. Define the project management framework
2. Set policies, processes and methodologies
3. Set limits of authority for decision-making

Area	Governance	
Domain	Establish	
Process ID	ES01	
Process Name	Define the project management framework	
Process Description	Clarify and maintain the governance framework for project management. Implement and maintain tools and techniques and authorities to manage projects in line with enterprise objectives.	
Process Purpose Statement	The purpose of *Define the project management framework* is to provide a consistent project management approach to enable enterprise governance requirements to be met, covering project management processes, organizational structures, roles and responsibilities, reliable and repeatable activities, and skills and competencies.	
OUTCOMES		
NUMBER	DESCRIPTION	
ES01-01	A project management framework is defined.	
ES01-02	A project management framework is followed.	
BASE PRACTICES (BPs)		
NUMBER	DESCRIPTION	SUPPORTS
ES01-BP1	Define a project management framework for IT investments.	ES01-01
ES01-BP2	Establish and maintain an IT project management framework	ES01-01 ES01-02
ES01-BP3	Establish Project Management Office	ES01-01

(*Continued*)

WORK PRODUCTS			
INPUTS			
NUMBER	DESCRIPTION		SUPPORTS
From corporate governance	Decision-making model		ES01-01 ES01-02
ES03-WP1	Authority limits		ES01-01 ES01-02
From corporate governance	Enterprise governance guiding principles		ES01-01 ES01-02
From corporate governance	Process architecture model		ES01-01
OUTPUTS			
NUMBER	DESCRIPTION	INPUT TO	SUPPORTS
ES01-WP1	Project management guidelines	All processes	ES01-01
ES01-WP2	PMO Charter	Project Manager	ES01-01
ES01-WP3	Definition of organizational structure and functions	All processes	ES01-01
ES01-WP4	Definition of project-related roles and responsibilities	All processes	ES01-01
ES01-WP5	Process capability assessments	All processes	ES01-02
ES01-WP6	Performance goals and metrics of process improvement tracking	All processes	ES01-02

REFERENCES	
DOCUMENT	SECTION
COBIT 5 Framework	APO01
ISO 21500:2012	3.4.2, 3.6
PMBOK 6th Edition	1.4
PRINCE2	6

Area	Governance	
Domain	Establish	
Process ID	ES02	
Process Name	Set policies, processes and methodologies	
Process Description	Develop, implement, and educate staff about the project management policy, supporting processes, tools and techniques.	
Process Purpose Statement	The purpose of *Set policies, processes and methodologies* is to develop project management policies, processes and methodologies and to ensure interested parties are knowledgeable of the guidance and trained in or made aware of aware of the guidance.	
OUTCOMES		
NUMBER	DESCRIPTION	
ES02-01	Policies, process and methodologies are defined and maintained.	
ES02-02	Policies, process and methodologies are known and followed.	
BASE PRACTICES (BPs)		
NUMBER	DESCRIPTION	SUPPORTS
ES02-BP1	Establish organizational project management policies	ES02-01
ES02-BP2	Standardize project initiation process	ES02-01
ES02-BP3	Standardize project plan development process	ES02-01
ES02-BP4	Standardize project scope planning process	ES02-01
ES02-BP5	Standardize project scope definition process	ES02-01
ES02-BP6	Standardize project activity definition process	ES02-01
ES02-BP7	Standardize project activity sequencing process	ES02-01
ES02-BP8	Standardize project activity duration estimating process	ES02-01
ES02-BP9	Standardize project schedule development process	ES02-01
ES02-BP10	Standardize project resource planning process	ES02-01
ES02-BP11	Standardize project cost estimating process	ES02-01
ES02-BP12	Standardize project cost budgeting process	ES02-01
ES02-BP13	Standardize project risk management planning process	ES02-01
ES02-BP14	Standardize project quality management planning process	ES02-01
ES02-BP15	Standardize project organizational planning process	ES02-01
ES02-BP16	Standardize project staff acquisition process	ES02-01
ES02-BP17	Standardize project communications planning process	ES02-01
ES02-BP18	Standardize project risk identification process	ES02-01
ES02-BP19	Standardize project quantitative risk analysis process	ES02-01
ES02-BP20	Standardize project qualitative risk analysis process	ES02-01
ES02-BP21	Standardize project risk response planning process	ES02-01
ES02-BP22	Standardize project procurement planning process	ES02-01
ES02-BP23	Standardize project solicitation planning process	ES02-01
ES02-BP24	Standardize project plan execution process	ES02-01
ES02-BP25	Standardize project quality assurance process	ES02-01

(*Continued*)

NUMBER	DESCRIPTION	SUPPORTS
ES02-BP26	Standardize project team development process	ES02-01
ES02-BP27	Standardize project information distribution process	ES02-01
ES02-BP28	Standardize project solicitation process	ES02-01
ES02-BP29	Standardize project source selection process	ES02-01
ES02-BP30	Standardize project contract administration process	ES02-01
ES02-BP31	Standardize project performance reporting process	ES02-01
ES02-BP32	Standardize project integrated change control process	ES02-01
ES02-BP33	Standardize project scope verification process	ES02-01
ES02-BP34	Standardize project scope change control process	ES02-01
ES02-BP35	Standardize project schedule control process	ES02-01
ES02-BP36	Standardize project cost control process	ES02-01
ES02-BP37	Standardize project quality control process	ES02-01
ES02-BP38	Standardize project risk monitoring and control process	ES02-01
ES02-BP39	Standardize project contract closeout process	ES02-01
ES02-BP40	Standardize project administrative closure process	ES02-01
ES02-BP41	Agree on core project management techniques	ES02-01
ES02-BP42	Ensure that standard processes are followed	ES02-02

WORK PRODUCTS

INPUTS

NUMBER	DESCRIPTION	SUPPORTS
Benchmarking system	Industry standards	ES02-01
Policy management system	Existing project policy and management system	ES02-01/2

OUTPUTS

NUMBER	DESCRIPTION	INPUT TO	SUPPORTS
ES02-WP1	Project management policy	All processes	ES02-01
ES02-WP2	Project management methodology	All management processes	ES02-01
ES02-WP3	Project management process definition documents	All management processes	ES02-01
ES02-WP4	Project documents	All management processes	ES02-02

REFERENCES

DOCUMENT	SECTION
ISO 21500:2012	4.3.1
OPM3	
PMBOK 6th Edition	Glossary

Area	Governance	
Domain	Establish	
Process ID	ES03	
Process Name	Set limits of authority for decision-making	
Process Description	Define distinct responsibilities for directing, managing and delivering the project; delegating authority and setting tolerances. Tolerances are the permissible deviation above and below a plan's target for project parameters without escalating the deviation to the next level of management. It also defines the level of formality for escalation.	
Process Purpose Statement	The purpose of *Set limits of authority for decision-making* is to: • Set out matters reserved for various roles in the project • Establish authority limits for each role.	
OUTCOMES		
NUMBER	DESCRIPTION	
ES03-01	Authority limits are established for each project role.	
ES03-02	Confirm authority limits with each role.	
BASE PRACTICES (BPs)		
NUMBER	DESCRIPTION	SUPPORTS
ES03-BP1	Establish the role of the project manager	ES03-01
ES03-BP2	Establish project manager competency processes	ES03-01
ES03-BP3	Commit resources for project initiation	ES03-02
ES03-BP4	Align capability with authority	ES03-01 ES03-02
ES03-BP5	Determine project complexity	ES03-01
ES03-BP6	Determine organizational capacity for change	ES03-01
ES03-BP7	Set tolerance for budget	ES03-01 ES03-02
ES03-BP8	Set tolerance for risk	ES03-01 ES03-02
ES03-BP9	Set tolerance for schedule	ES03-01 ES03-02
ES03-BP10	Set tolerance for quality	ES03-01 ES03-02
ES03-BP11	Set tolerance for scope	ES03-01 ES03-02
ES03-BP12	Set tolerance for benefits	ES03-01 ES03-02

(*Continued*)

WORK PRODUCTS			
INPUTS			
NUMBER	DESCRIPTION		SUPPORTS
From Board of Directors	Delegation of authority policy		ES03-01
Human resource management system	Organization charts		ES03-01
Enterprise risk management system	Enterprise risk policy		ES03-01
Enterprise risk management system	Risk appetite		ES03-01
Enterprise quality management system	Enterprise quality policy		ES03-01
Financial management system	Project budget		ES03-01
OUTPUTS			
NUMBER	DESCRIPTION	INPUT TO	SUPPORTS
ES03-WP1	Authority limits	All processes	ES03-01 ES03-02
ES03-WP2	Budget tolerance	PL01-01	ES03-01
ES03-WP3	Risk tolerance	PL01-01	ES03-01
ES03-WP4	Schedule tolerance	PL01-01	ES03-01
ES03-WP5	Quality tolerance	PL01-01	ES03-01
ES03-WP6	Scope tolerance	PL01-01	ES03-01
ES03-WP7	Benefits tolerance	PL01-01	ES03-01

REFERENCES	
DOCUMENT	SECTION
ISO 21500:2012	3.6, 3.11
OPM3	Appendix F
PMBOK 6th Edition	9.5.2.1
PRINCE2	2.5 Manage by Exception

Monitor Domain

1. Ensure project benefits
2. Ensure risk optimization
3. Ensure resource optimization

Area	Governance
Domain	Monitor
Process ID	MO01
Process Name	Ensure project benefits
Process Description	Optimize the value contribution to the business from project management processes resulting from the investments made by acceptable costs.
Process Purpose Statement	The purpose of *Ensure project benefits* is to secure optimal business value from projects and to provide a reliable and accurate statement of costs and benefits so that business needs are met.
OUTCOMES	
NUMBER	DESCRIPTION
MO01-01	The enterprise is securing optimal value from projects.
MO01-02	Optimal value is derived from investments through effective value management practices.
MO01-03	Individual projects contribute optimal value.

BASE PRACTICES (BPs)		
NUMBER	DESCRIPTION	SUPPORTS
M001-BP1	Include strategic goals into project objectives or Ensure project objectives support strategic goals	M001-01
M001-BP2	Optimize portfolio management	M001-01
M001-BP3	Align projects	M001-01
M001-BP4	Optimize project strategic alignment	M001-01
M001-BP5	Know inter-project plan	M001-01
M001-BP6	Review projects against "Continue, Defer or Terminate" criteria	M001-01
M001-BP7	Prioritize projects	M001-01
M001-BP8	Link performance measurement to project life cycles	M001-01
M001-BP9	Select projects based on benefits and organizational business value	M001-01

WORK PRODUCTS		
INPUTS		
NUMBER	DESCRIPTION	SUPPORTS
M001-WP1	Business case methodology	M001-01
M001-WP2	Benefits realization methodology	M001-01

OUTPUTS			
NUMBER	DESCRIPTION	INPUT TO	SUPPORTS
M001-WP1	Business case document	M001	M001-01
M001-WP2	Benefits review plan	All Governance	M001-02 M001-03

REFERENCES	
DOCUMENT	SECTION
COBIT 5 Framework	EDM02
ISO 21500:2012	3.1, 3.4.1, 3.4.3
OPM3	Appendix F
PRINCE2	1.5, 3.1.2, 4.3.3, A.1

Area	Governance	
Domain	Monitor	
Process ID	MO02	
Process Name	Ensure risk optimization	
Process Description	Ensure that the enterprise's risk appetite and tolerance are recognized, articulated and communicated, and risk to enterprise value related to projects is identified and managed.	
Process Purpose Statement	The purpose of *Ensure risk optimization* is to ensure that enterprise risk for projects does not exceed risk appetite and risk tolerance, the impact of project risk to the enterprise is identified and managed, and the potential for compliance failures is minimized.	
OUTCOMES		
NUMBER	DESCRIPTION	
MO02-01	Risk thresholds are defined and communicated, and project-related risk is known.	
MO02-02	The enterprise is managing critical project-related enterprise risk effectively and efficiently.	
MO02-03	Project-related risk does not exceed risk appetite and tolerance, and the impact of project risk to enterprise value is identified and managed.	
BASE PRACTICES (BPs)		
NUMBER	DESCRIPTION	SUPPORTS
MO02-BP1	Establish risk management.	MO02-02
MO02-BP2	Evaluate risk management.	MO02-01 MO02-02
MO02-BP3	Direct risk management.	MO02-02 MO02-03
MO02-BP4	Monitor risk management.	MO02-03
MO02-BP5	Encourage risk taking to realize opportunities.	MO02-01 MO02-03
WORK PRODUCTS		
INPUTS		
NUMBER	DESCRIPTION	SUPPORTS
Enterprise risk management system[1]	Risk appetite	MO02-01 MO02-03
ES03-WP1	Authority limits	MO02-01 MO02-02
ES03-WP2	Risk tolerance	MO02-01

(*Continued*)

[1] Refer to ISO 31000.

OUTPUTS			
NUMBER	DESCRIPTION	INPUT TO	SUPPORTS
MO02-WP1	Risk management policy	All processes	MO02-01
MO02-WP2	Risk appetite guidance	PL12	MO02-01
MO02-WP3	Approved project risk tolerance levels	PL12	MO02-01
MO02-WP4	Evaluation of risk management activities	PL12	MO02-02
MO02-WP5	Approved process for measuring project risk management	PL12	MO02-02 MO02-03
MO02-WP6	Project risk management issues for the Project Steering Committee/Project Governance body	PL12	MO02-02 MO02-03

REFERENCES	
DOCUMENT	SECTION
COBIT 5 Framework	EDM03
ISO 21500:2012	3.4.1, 3.5.2, 3.6, 3.11, 4.1
OPM3	Appendix F
PMBOK 6th Edition	11
PRINCE2	3.5

Area	Governance
Domain	Monitor
Process ID	MO03
Process Name	Ensure resource optimization
Process Description	Ensure that adequate and sufficient project-related capabilities; that is, people, process and technology are available to support objectives effectively at optimal cost.
Process Purpose Statement	The purpose of *Ensure resource optimization* is to ensure that resource needs for project management of the project are met in optimal manner, project costs are optimized, and there is an increased likelihood of benefit realization and readiness for future change.

OUTCOMES	
NUMBER	DESCRIPTION
MO03-01	The project resource needs of the enterprise are met with optimal capabilities.
MO03-02	Project resources are allocated to meet enterprise priorities within project budget constraints.
MO03-03	Optimal use of project resources is achieved throughout the full economic life cycle to each resource.

BASE PRACTICES (BPs)		
NUMBER	DESCRIPTION	SUPPORTS
MO03-BP1	Evaluate resource requirements.	MO03-01
MO03-BP2	Staff projects with competent resources.	MO03-02
MO03-BP3	Manage project resource pool.	MO03-02 MO03-03
MO03-BP4	Monitor resources.	MO03-03

WORK PRODUCTS		
INPUTS		
NUMBER	DESCRIPTION	SUPPORTS
ES02-WP3	Project management process definition documents	MO03-01 MO03-02 MO03-03
From human resources management system	Skills development plans	MO03-01

OUTPUTS			
NUMBER	DESCRIPTION	INPUT TO	SUPPORTS
MO03-WP1	Resource availability	IN03	MO03-01
MO03-WP2	Approved resource plan	IN03	MO03-02

(*Continued*)

NUMBER	DESCRIPTION	INPUT TO	SUPPORTS
MO03-WP3	Feedback on allocation and effectiveness of resources and capabilities	IN03	MO03-03
MO03-WP4	Remedial actions to address resource management gaps	IN03	MO03-02 MO03-03

REFERENCES

DOCUMENT	SECTION
COBIT 5 Framework	EDM04
ISO 21500:2012	3.5.1, 3.11, 4.1
OPM3	Appendix F
PMBOK 6th Edition	6.5.2.3
PRINCE2	5.6

Initiate Domain

1. Develop project charter
2. Identify stakeholders
3. Establish project team

Area	Management	
Domain	Initiate	
Process ID	IN01	
Process Name	Develop project charter	
Process Description	Develop a document that formally authorizes the existence of a project and provides the project manager the authority to apply organizational resources to project activities.	
Process Purpose Statement	The purpose of *Develop project charter* is: • To formally authorize a project or a new project phase; • To identify the project manager and the appropriate project manager responsibilities and authorities; and, • To document the business needs, project objectives, expected deliverables and the economic aspects of the project.	
OUTCOMES		
NUMBER	DESCRIPTION	
IN01-01	Well-defined project charter. Change start to charter—start is when the business case is approved, and the mandate is issued	
IN01-02	Well-defined project boundaries.	
IN01-03	Formal record of the project as part of the portfolio.	
BASE PRACTICES (BPs)		
NUMBER	DESCRIPTION	SUPPORTS
IN01-BP1	Adhere to inter-project rules of conduct	IN01-01
IN01-BP2	Assess inputs using expert judgement	IN01-02
IN01-BP3	Facilitate sessions to brainstorm project activities	IN01-02
IN01-BP4	Author the project charter	IN01-03

(*Continued*)

WORK PRODUCTS			
INPUTS			
NUMBER	DESCRIPTION		SUPPORTS
Outside project management	Contract or Agreement		IN01-01
Outside project management	Enterprise environmental factors		IN01-02
Management	Project statement of work		IN01-02 IN01-03
MO01-WP3	Business case document		IN01-01 IN-01-03
Outside project management	Organizational process assets—these primarily are internal		IN01-03
OUTPUTS			
NUMBER	DESCRIPTION	INPUT TO	SUPPORTS
IN01-WP1	Project charter	Develop project management plan, Plan scope management Plan resource management	IN01-01

REFERENCES	
DOCUMENT	SECTION
COBIT 5 Framework	AP006 BAI01
ISO 21500:2012	4.3.2
OPM3	Appendix F
PMBOK 6th Edition	4.1.3
PRINCE2	4.1.1

Area	Management		
Domain	Initiate		
Process ID	IN02		
Process Name	Identify stakeholders		
Process Description	Ensure that internal and external stakeholders are identified, and their authority, involvement and interest is documented.		
Process Purpose Statement	The purpose of *Identify stakeholders* is to determine the individuals, groups or organizations affected by, or affecting, the project and to document relevant information regarding their authority, interest and involvement.		
OUTCOMES			
NUMBER	DESCRIPTION		
IN02-01	Stakeholders and their needs are identified.		
BASE PRACTICES (BPs)			
NUMBER	DESCRIPTION		SUPPORTS
IN02-BP1	Identify stakeholders.		IN02-01
IN02-BP2	Establish strong sponsorship.		IN02-01
IN02-BP3	Evaluate stakeholder reporting requirements.		IN02-01
IN02-BP4	Perform stakeholder analysis.		IN02-01
IN02-BP5	Maintain stakeholder register.		IN02-01
WORK PRODUCTS			
INPUTS			
NUMBER	DESCRIPTION		SUPPORTS
IN01-WP1	Project charter		IN02-01
PL06-WP2	Project organization chart		IN02-01
OUTPUTS			
NUMBER	DESCRIPTION	INPUT TO	SUPPORTS
IN02-WP1	Stakeholder register	PL06, PL16, IM02	IN02-01
IN02-WP2	Escalation guidelines	All processes	IN02-01
IN02-WP3	Risk analysis and risk profile reports for stakeholders	IN02, IM02, IM04, CO08	IN02-01

(*Continued*)

REFERENCES	
DOCUMENT	SECTION
AA1000	4.1
COBIT 5 Framework	EDM05, BAI01
ISO 21500:2012	4.3.9
OPM3	Appendix F
PMBOK 6th Edition	13.1
PRINCE2	3.2

Area	Management		
Domain	Initiate		
Process ID	IN03		
Process Name	Establish project team		
Process Description	Determine how many and when project team members are needed considering skills and expertise, and when the project will release them from the project. When team members are not available within the organization, acquire the resources from outside.		
Process Purpose Statement	The purpose of *Establish project team* is to acquire the human resources needed to complete the project.		
OUTCOMES			
NUMBER	DESCRIPTION		
IN03-01	Project team is competent and able to deliver on the project plans.		
BASE PRACTICES (BPs)			
NUMBER	DESCRIPTION		SUPPORTS
IN03-BP1	Define project team structure.		IN03-01
IN03-BP2	Establish an effective project team by assembling appropriate members, establishing swift trust, and establishing common goals and effectiveness measures.		IN03-01
IN03-BP3	Define team member performance targets.		IN03-01
WORK PRODUCTS			
INPUTS			
NUMBER	DESCRIPTION		SUPPORTS
PL05-WP1	Resource requirements		IN03-01 IN03-BP1
PL06-WP2	Project organization chart		IN03-01 IN03-BP1
MO03-WP1	Resource availability		IN03-01 IN03-BP2
PL06-WP1	Role descriptions		IN03-01 IN03-BP2
OUTPUTS			
NUMBER	DESCRIPTION	INPUT TO	SUPPORTS
IN03-WP1	Team performance	Performance measurement systems	IN03-01 IN03-BP2
IN03-WP2	Team appraisal	Performance measurement systems	IN03-01 IN03-BP2

(*Continued*)

REFERENCES	
DOCUMENT	SECTION
COBIT 5 Framework	BAI01
ISO 21500:2012	4.3.15
OPM3	Appendix F
PRINCE2	3.2

Plan Domain

1. Develop project plans
2. Define scope
3. Create work breakdown structure
4. Define activities
5. Estimate resources
6. Define project organization
7. Sequence activities
8. Estimate activity durations
9. Develop schedule
10. Estimate costs
11. Develop budget
12. Identify risks
13. Assess risks
14. Plan quality
15. Plan procurements
16. Plan communications

Area	Management
Domain	Plan
Process ID	PL01
Process Name	Develop project plans
Process Description	Initiate projects in a coordinated way.
Process Purpose Statement	The purpose of *Develop project plans* is to document the following: —why the project is being undertaken; —what will be provided and by whom; —how it will be provided; —what it will cost; —how the project will be implemented, controlled and closed
OUTCOMES	
NUMBER	DESCRIPTION
PL01-01	Project plans are likely to achieve the expected outcomes.

(*Continued*)

BASE PRACTICES (BPs)		
NUMBER	DESCRIPTION	SUPPORTS
PL01-BP1	Adapt project management processes.	PL01-01
PL01-BP2	Integrate project management methods.	PL01-01
PL01-BP3	Use standard planning baseline	PL01-01
PL01-BP4	Set project objectives.	PL01-01
PL01-BP5	Quantify specifications.	PL01-01
PL01-BP6	Establish mathematical models for planning.	PL01-01

WORK PRODUCTS		
INPUTS		
NUMBER	DESCRIPTION	SUPPORTS
IN01-WP1	Project charter	PL01-01
From project management processes	Subsidiary plans	PL01-01
CL02-WP1	Lessons learned from previous projects	PL01-01
MO01-WP3	Business case document	PL01-01
Change management system[2]	Approved changes	PL01-01
Financial management system	Cost baseline	PL01-01
Time or schedule management system	Schedule baseline	PL01-01
Strategic planning process	Enterprise organizational factors	PL01-01
Asset management system	Organizational process assets	PL01-01

OUTPUTS			
NUMBER	DESCRIPTION	INPUT TO	SUPPORTS
PL01-WP1	Project plans	IM01-01, CO02-01, CL02-01	PL01-01
PL01-WP2	Project management plan	All processes	PL01-01

REFERENCES	
DOCUMENT	SECTION
COBIT 5 Framework	BAI01
ISO/IEC 15289:2017	7, 10
ISO 21500:2012	4.3.3
OPM3	Appendix F
PMBOK 6th Edition	4.2
PRINCE2	3.4

[2] ISO 20000 provides an example of an information technology change management system.

Area	Management		
Domain	Plan		
Process ID	PL02		
Process Name	Define scope		
Process Description	Define what the project will contribute to the strategic goals of the organization as the basis for project decisions and communication of the organizational benefits.		
Process Purpose Statement	The purpose of *Define scope* is to achieve clarity of the project scope, including objectives, deliverables, requirements, constraints and boundaries by defining the end state of the project or vision. or vision.		
OUTCOMES			
NUMBER	DESCRIPTION		
PL02-01	Definition of project end state or vision.		
BASE PRACTICES (BPs)			
NUMBER	DESCRIPTION		SUPPORTS
PL02-BP1	Perform product analysis		PL02-01
PL02-BP2	Generate alternatives		PL02-01
PL02-BP3	Determine project scope		PL02-01
WORK PRODUCTS			
INPUTS			
NUMBER	DESCRIPTION		SUPPORTS
IN01-WP1	Project charter		PL02-01
Change management system	Approved changes		CO02-01
OUTPUTS			
NUMBER	DESCRIPTION	INPUT TO	SUPPORTS
PL02-WP1	Scope statement	CO03	PL02-01
PL02-WP2	Requirements document	PL03	PL02-01
REFERENCES			
DOCUMENT		SECTION	
COBIT 5 Framework		BAI01	
ISO 21500:2012		4.3.11	
OPM3		Appendix F	
PMBOK 6th Edition		5.3	
PRINCE2		1.5	

Area	Management		
Domain	Plan		
Process ID	PL03		
Process Name	Create a work breakdown structure		
Process Description	Provide a framework for decomposing and describing the overall project into manageable and detailed pieces of work by phase, deliverable, discipline or location.		
Process Purpose Statement	The purpose of *Create work breakdown structure* is to provide a hierarchical decomposition framework for presenting the work that needs to be completed in order to achieve the project objectives		
OUTCOMES			
NUMBER	DESCRIPTION		
PL03-01	Hierarchical decomposition of project work.		
BASE PRACTICES (BPs)			
NUMBER	DESCRIPTION		SUPPORTS
PL03-BP1	Define hierarchical work breakdown structure.		PL03-01
PL03-BP2	Create work packages.		PL03-01
WORK PRODUCTS			
INPUTS			
NUMBER	DESCRIPTION		SUPPORTS
IN01-WP1	Project plans		PL03-01
PL02-WP2	Requirements		PL03-01
Project Planning	Project scope statement		PL03-01
Change management system I am confused as to why some have numbers and others do not	Approved changes		PL03-01
OUTPUTS			
NUMBER	DESCRIPTION	INPUT TO	SUPPORTS
PL03-WP1	Work breakdown structure	PL04, 06, 10,11 CO03	PL03-01
PL03-WP2	Work breakdown structure dictionary	PL04	PL03-01
REFERENCES			
DOCUMENT		SECTION	
COBIT 5 Framework		BAI01	
ISO 21500:2012		4.3.12	
PMBOK 6th Edition		5.4	
PRINCE2		4.4.3, A.26	

Area	Management		
Domain	Plan		
Process ID	PL04		
Process Name	Define activities		
Process Description	Identifies, defines and documents all activities for scheduling and monitoring based on the work breakdown structure. Provides basis for planning, implementing, controlling and closing project work.		
Process Purpose Statement	The purpose of **Define activities** is to identify, define and document all the activities that should be in the schedule and performed in order to achieve the project objectives.		
OUTCOMES			
NUMBER	DESCRIPTION		
PL04-01	Definition of work activities. Or activity list		
BASE PRACTICES (BPs)			
NUMBER	DESCRIPTION		SUPPORTS
PL04-BP1	Identify work activities from the work breakdown structure.		PL04-01
PL04-BP2	Define work activities.		PL04-01
PL04-BP3	Document work activities in a list.		PL04-01
WORK PRODUCTS			
INPUTS			
NUMBER	DESCRIPTION		SUPPORTS
PL03-WP1	Work breakdown structure		PL04-01
PL03-WP2	Work breakdown structure dictionary		PL04-01
PL01-WP1	Project plan		PL04-01
Change management system	Approved changes		PL04-01
OUTPUTS			
NUMBER	DESCRIPTION	INPUT TO	SUPPORTS
PL04-WP1	Activity list	PL05, 07, 08,09,10 CO03	PL04-01
PL04-WP2	Activity attributes	PL05, 07, 08, 09	PL04-01
PL04-WP3	Milestone list	PL07	PL04-01
REFERENCES			
DOCUMENT		SECTION	
COBIT 5 Framework		BAI01	
ISO 21500:2012		4.3.13	
PMBOK 6th Edition		6.2	

Area	Management		
Domain	Plan		
Process ID	PL05		
Process Name	Estimate resources		
Process Description	Record resource units and engagement start and end.		
Process Purpose Statement	The purpose of *Estimate resources* is to determine the resources needed for each activity in the activity list. Resources may include people, facilities, equipment, materials, infrastructure and tools		
OUTCOMES			
NUMBER	DESCRIPTION		
PL05-01	Identified type, quantity and characteristics of resources required to complete the activity.		
BASE PRACTICES (BPs)			
NUMBER	DESCRIPTION		SUPPORTS
PL05-BP1	Identify risks		PL05-01
PL05-BP2	Estimate activity effort		PL05-01
PL05-BP3	Create a resource breakdown structure		PL05-01
PL05-BP4	Document activity resource requirements		PL05-01
PL05-BP5	Update project plan		PL05-01
WORK PRODUCTS			
INPUTS			
NUMBER	DESCRIPTION		SUPPORTS
PL04-WP1	Activity list		PL05-01
PL04-WP2	Activity attributes		PL05-01
	Activity cost estimates		PL05-01
PL01-WP1	Project plans		PL05-01
Schedule management system	Schedule		PL05-01
Change management system	Approved changes		PL05-01
Time or schedule management system	Resource calendars		PL05-01
PL12-WP1	Risk register		PL05-01
OUTPUTS			
NUMBER	DESCRIPTION	INPUT TO	SUPPORTS
PL05-WP1	Resource requirements	IN03, PL06, 08, 09, 15 CO04	PL05-01
PL05-WP2	Resource plan	IM03, CO05	PL05-01

(*Continued*)

REFERENCES	
DOCUMENT	SECTION
COBIT 5 Framework	BAI01
ISO 21500:2012	4.3.16
PMBOK 6th Edition	9.2
PRINCE2	5.6

Area	Management		
Domain	Plan		
Process ID	PL06		
Process Name	Define project organization		
Process Description	Define project organizational structure including the identification of all team members and other parties directly involved in the project work. Includes the assignment of project responsibilities and authorities at the work breakdown structure level to perform the approved work, manage progress and allocate resources.		
Process Purpose Statement	The purpose of *Define project organization* is to secure all needed commitments from all the parties involved in a project. Roles, responsibilities and authorities that are relevant to the project should be defined in accordance with the project's nature and complexity and should consider the performing organization's existing policies.		
OUTCOMES			
NUMBER	DESCRIPTION		
PL06-01	Required team to execute the project plan and deliver services or products.		
BASE PRACTICES (BPs)			
NUMBER	DESCRIPTION		SUPPORTS
PL06-BP1	Define project team structure		PL06-01
PL06-BP2	Use teamwork		PL06-01
WORK PRODUCTS			
INPUTS			
NUMBER	DESCRIPTION		SUPPORTS
PL01-WP1	Project plans		PL06-01
PL03-WP1	Work breakdown structure		PL06-01
PL05-WP1	Resource requirements		PL06-01
IN02-WP1	Stakeholder register		PL06-01
Change management system	Approved changes		PL06-01
OUTPUTS			
NUMBER	DESCRIPTION	INPUT TO	SUPPORTS
PL06-WP1	Role descriptions	IM03, CO05, PL16	PL06-01
PL06-WP2	Project organization chart	CO05, IN02	PL06-01

(*Continued*)

REFERENCES	
DOCUMENT	SECTION
COBIT 5 Framework	BAI01
ISO 21500:2012	4.3.17
OPM3	Appendix F
PMBOK 6th Edition	9.1.3

Area	Management		
Domain	Plan		
Process ID	PL07		
Process Name	Sequence activities		
Process Description	Sequence activities logically, which includes proper precedence relationships and appropriate leads, lags, constraints, interdependencies and external dependencies to support development of a realistic and achievable project schedule.		
Process Purpose Statement	The purpose of *Sequence activities* is to identify and document the logical relationships between project activities.		
OUTCOMES			
NUMBER	DESCRIPTION		
PL07-01	Logical sequence of project activities.		
BASE PRACTICES (BPs)			
NUMBER	DESCRIPTION		SUPPORTS
PL07-BP1	Determine activity relationships.		PL07-01
PL07-BP2	Sequence activities.		PL07-01
PL07-BP3	Determine the critical path.		PL07-01
WORK PRODUCTS			
INPUTS			
NUMBER	DESCRIPTION		SUPPORTS
PL04-WP1	Activity list		PL07-01
PL04-WP2	Activity attributes		PL07-01
PL04-WP3	Milestone list		PL07-01
Project Planning	Project scope statement		PL07-01
Change management system	Approved changes		PL07-01
OUTPUTS			
NUMBER	DESCRIPTION	INPUT TO	SUPPORTS
PL07-WP1	Activity sequence	PL09	PL07-01
REFERENCES			
DOCUMENT		SECTION	
COBIT 5 Framework		BAI01	
ISO 21500:2012		4.3.21	
PMBOK 6th Edition		6.3	

Area	Management		
Domain	Plan		
Process ID	PL08		
Process Name	Estimate activity durations		
Process Description	Estimate activity durations based on quantity and type of resource needed, relationship between activities, capacities, capabilities, schedules, activity knowledge and administrative overhead.		
Process Purpose Statement	The purpose of *Estimate activity durations* is to estimate the time required to complete each activity in the project.		
OUTCOMES			
NUMBER	DESCRIPTION		
PL08-01	Amount of time each activity will take to complete.		
BASE PRACTICES (BPs)			
NUMBER	DESCRIPTION		SUPPORTS
PL08-BP1	Estimate project task duration		PL08-01
PL08-BP2	Identify risks		PL08-01
WORK PRODUCTS			
INPUTS			
NUMBER	DESCRIPTION		SUPPORTS
Project Planning	Project scope statement		PL08-01
PL04-WP1	Activity list		PL08-01
PL04-WP2	Activity attributes		PL08-01
PL05-WP1	Resource requirements		PL08-01
PL04-WP1	Resource breakdown structure		PL08-01
Knowledge management system	Historical data		PL08-01
Knowledge management system Include with that above	Industry standards		PL08-01
PL12-WP1	Risk register		PL08-01
Change management system	Approved changes		PL08-01
OUTPUTS			
NUMBER	DESCRIPTION	INPUT TO	SUPPORTS
PL08-WP1	Activity duration estimates	PL09	PL08-01
REFERENCES			
DOCUMENT		SECTION	
COBIT 5 Framework		BAI01	
ISO 21500:2012		4.3.22	
OPM3		Appendix F	
PMBOK 6th Edition		6.4	

Area	Management		
Domain	Plan		
Process ID	PL09		
Process Name	Develop schedule		
Process Description	Activities are scheduled in a logical sequence identifying durations, milestones and interdependencies.		
Process Purpose Statement	The purpose of *Develop schedule* is to calculate the start and end times of the project activities and to establish the overall project schedule baseline.		
OUTCOMES			
NUMBER	DESCRIPTION		
PL09-01	Accurate schedule of activities as overall project baseline.		
BASE PRACTICES (BPs)			
NUMBER	DESCRIPTION		SUPPORTS
PL09-BP1	Develop schedule		PL09-01
PL09-BP2	Update schedule throughout project		PL09-01
PL09-BP3	Identify risks		PL09-01
WORK PRODUCTS			
INPUTS			
NUMBER	DESCRIPTION		SUPPORTS
PL07-WP1	Activity sequence		PL09-01
PL04-WP1	Activity list		PL09-01
PL04-WP2	Activity attributes		PL09-01
PL08-WP1	Activity duration estimates		PL09-01
PL05-WP1	Resource requirements		PL09-01
Strategic planning	Schedule constraints		PL09-01
Project planning	Project scope statement		PL09-01
PL12-WP1	Risk register		PL09-01
Change management system	Approved changes		PL09-01
OUTPUTS			
NUMBER	DESCRIPTION	INPUT TO	SUPPORTS
PL09-WP1	Schedule	IN09, CO06	PL09-01
REFERENCES			
DOCUMENT		SECTION	
COBIT 5 Framework		BAI01	
ISO 21500:2012		4.3.23	
PMBOK 65th Edition		6.5	

Area	Management		
Domain	Plan		
Process ID	PL10		
Process Name	Estimate costs		
Process Description	Estimate and state costs in terms of currency or a standard unit of measure, such as number of equipment hours and labor hours.		
Process Purpose Statement	The purpose of *Estimate costs* is to obtain an approximation of the costs needed to complete each project activity and for the project as a whole.		
OUTCOMES			
NUMBER	DESCRIPTION		
PL10-01	Accurate cost estimates using an accepted unit of measurement.		
BASE PRACTICES (BPs)			
NUMBER	DESCRIPTION		SUPPORTS
PL10-BP1	Determine unit of measurement for estimation.		PL10-01
PL10-BP2	Estimate costs based on some valuation.		PL10-01
PL10-BP3	Estimate reserves or contingencies for risk or uncertainties.		PL10-01
WORK PRODUCTS			
INPUTS			
NUMBER	DESCRIPTION		SUPPORTS
PL03-WP1	Work breakdown structure		PL10-01
PL04-WP1	Activity list		PL10-01
PL01-WP1	Project plans		PL10-01
Change management system	Approved changes		PL10-01
OUTPUTS			
NUMBER	DESCRIPTION	INPUT TO	SUPPORTS
PL10-WP1	Cost estimates	PL11	PL10-01
REFERENCES			
DOCUMENT		SECTION	
COBIT 5 Framework		BAI01	
ISO 21500:2012		4.3.25	
PMBOK 6th Edition		7.2	

Area	Management		
Domain	Plan		
Process ID	PL11		
Process Name	Develop budget		
Process Description	Assign budget to scheduled work and use it for comparison to actual performance.		
Process Purpose Statement	The purpose of *Develop budget* is to distribute the project's budget to the appropriate levels of the work breakdown structure at the correct time.		
OUTCOMES			
NUMBER	DESCRIPTION		
PL11-01	Realistic budget tied to an established scope of work.		
BASE PRACTICES (BPs)			
NUMBER	DESCRIPTION		SUPPORTS
PL11-BP1	Set objective measures of cost performance.		PL11-01
PL11-BP2	Establish when will be expended.		PL11-01
PL11-BP3	Assess cost performance using an established method.		PL11-01
PL11-BP4	Create reserves or contingencies for management control purposes or risk coverage.		PL11-01
WORK PRODUCTS			
INPUTS			
NUMBER	DESCRIPTION		SUPPORTS
PL03-WP1	Work breakdown structure		PL11-01
PL10-WP1	Cost estimates		PL11-01
PL09-WP1	Schedule		PL11-01
PL01-WP1	Project plans		PL11-01
Change management system	Approved changes		PL11-01
OUTPUTS			
NUMBER	DESCRIPTION	INPUT TO	SUPPORTS
PL11-WP1	Budget	CO07	PL11-01
REFERENCES			
DOCUMENT		SECTION	
COBIT 5 Framework		BAI01	
ISO 21500:2012		4.3.26	
PMBOK 6th Edition		7.3	

Area	Management		
Domain	Plan		
Process ID	PL12		
Process Name	Identify risks		
Process Description	Continually identify project-related risk within levels of tolerance set by project steering committee or project Governance Body.		
Process Purpose Statement	The purpose of *Identify risks* is to determine potential risk events and their characteristics that, if they occur, may have a positive or negative impact on the project objectives.		
OUTCOMES			
NUMBER	DESCRIPTION		
PL12-01	Project-related risk is identified.		
PL12-02	A current and complete project risk profile exists.		
BASE PRACTICES (BPs)			
NUMBER	DESCRIPTION		SUPPORTS
PL12-BP1	Identify and collect relevant data to enable project risk identification		PL12-01
PL12-BP2	Maintain an inventory of known project risks.		PL12-02
PL12-BP3	Provide information on identified project risks.		PL12-02
WORK PRODUCTS			
INPUTS			
NUMBER	DESCRIPTION		SUPPORTS
From enterprise risk management system	Potential project threats		PL12-01
PL01-WP1	Project plans		PL12-01
From enterprise risk management system	Risk profile format		PL12-02
OUTPUTS			
NUMBER	DESCRIPTION	INPUT TO	SUPPORTS
PL12-WP1	Risk register	PL13, PL15, IM04, CO01, CO08, CL02	PL12-01
REFERENCES			
DOCUMENT		SECTION	
COBIT 5 Framework		AP012 BAI01	
ISO 21500:2012		4.3.28	
PMBOK 6th Edition		11.2	

Area	Management		
Domain	Plan		
Process ID	PL13		
Process Name	Assess risks		
Process Description	Continually assess risks to ensure risk is within risk tolerance level given to project from project steering committee/project Governing Body.		
Process Purpose Statement	The purpose of *Assess risks* is to measure and prioritize the risks for further action or acceptance.		
OUTCOMES			
NUMBER	DESCRIPTION		
PL13-01	Project-related risk and analyzed.		
PL13-02	Risk assessment activities are done effectively.		
BASE PRACTICES (BPs)			
NUMBER	DESCRIPTION		SUPPORTS
PL13-BP1	Assess risks to develop useful information to support risk decisions.		PL13-01
PL13-BP2	Provide information on the current state of project-related threats and opportunities in a timely manner to all required stakeholders for appropriate response.		PL13-01 PL13-02
WORK PRODUCTS			
INPUTS			
NUMBER	DESCRIPTION		SUPPORTS
PL12-WP1	Risk register		PL13-01
PL01-WP1	Project plans		PL13-01
OUTPUTS			
NUMBER	DESCRIPTION	INPUT TO	SUPPORTS
PL13-WP1	Prioritized risks	CO08	PL13-01
REFERENCES			
DOCUMENT		SECTION	
COBIT 5 Framework		APO12 BAI01	
ISO 21500:2012		4.3.29	
OPM3		Appendix F	
PMBOK 6th Edition		11.3, 11.4	
PRINCE2		3.5	

Area	Management		
Domain	Plan		
Process ID	PL14		
Process Name	Plan quality		
Process Description	Define project quality requirements in all activities, procedures, and the related outcomes, including controls.		
Process Purpose Statement	The purpose of *Plan quality* is to determine the quality requirements and standards that will be applicable to the project, the deliverables of the project and how the requirements and standards will be met based on the project objectives.		
OUTCOMES			
NUMBER	DESCRIPTION		
PL14-01	Quality requirements are defined for the project.		
BASE PRACTICES (BPs)			
NUMBER	DESCRIPTION		SUPPORTS
PL14-BP1	Define project quality standards, practices, and procedures		PL14-01
WORK PRODUCTS			
INPUTS			
NUMBER	DESCRIPTION		SUPPORTS
PL01-WP1	Project plans		PL14-01
Quality management system[3]	Quality requirements		PL14-01
Enterprise quality management system	Quality policy		PL14-01
Change management system	Approved changes		PL14-01
OUTPUTS			
NUMBER	DESCRIPTION	INPUT TO	SUPPORTS
PL14-WP1	Quality plan	IM05, CO09	PL14-01
PL14-WP2	Quality roles, responsibilities and decision rights	IM05, CO09	PL14-01
PL14-WP3	Quality management standards	All Management Processes	PL14-01
PL14-WP4	Process improvement plan	Internal to process	PL14-01
PL14-WP5	Customer requirements for quality management	IM05, CO09	PL14-01
PL14-WP6	Acceptance criteria	IM05, CO09	PL14-01
REFERENCES			
DOCUMENT		SECTION	
COBIT 5 Framework		APO11 BAI01	
ISO 21500:2012		4.3.32	
PMBOK 6th Edition		8.1	
PRINCE2		3.3	

[3] Refer to ISO 9001:2015.

Area	Management	
Domain	Plan	
Process ID	PL15	
Process Name	Plan procurements	
Process Description	Plan and document the procurement strategy to facilitate procurement decision-making and to develop procurement specifications and requirements.	
Process Purpose Statement	The purpose of *Plan procurements* is to plan and document the procurement strategy and overall process properly before procurements are initiated.	
OUTCOMES		
NUMBER	DESCRIPTION	
PL15-01	Make-or-buy decision	
PL15-02	Strategy for procuring resources	
BASE PRACTICES (BPs)		
NUMBER	DESCRIPTION	SUPPORTS
PL15-BP1	Perform make-or-buy analysis	PL15-01
PL15-BP2	Perform market research	PL15-02
PL15-BP3	Identification of what to acquire, how to acquire it, how much is needed, and when to acquire it	PL15-02
WORK PRODUCTS		
INPUTS		
NUMBER	DESCRIPTION	SUPPORTS
PL01-WP1	Project plans	PL15-01 PL15-03
Capacity management system,[4] Human resource management system	In-house capacity and capability	PL15-01
Supplier management system[5]	Existing contracts	PL15-02
PL05-WP1	Resource requirements	PL15-01 PL15-02
PL09-WP1	Schedule	PL15-02
PL12-WP1	Risk register	PL15-01 PL15-02

(*Continued*)

[4] ISO 20000 gives a good example of IT capacity management system.

[5] ISO 20000 gives a good example of IT supplier management system.

OUTPUTS			
NUMBER	DESCRIPTION	INPUT TO	SUPPORTS
PL15-WP1	Procurement plan	IM06	PL15-02
PL15-WP2	Preferred supplier list	Supplier management system	PL15-02
PL15-WP3	Make-or-buy decision list	IM06	PL15-01
PL15-WP4	Change requests	Change management system	PL15-01 PL15-02

REFERENCES	
DOCUMENT	SECTION
COBIT 5 Framework	BAI01
ISO 21500:2012	4.3.35
PMBOK 6th Edition	12.1

Area	Management		
Domain	Plan		
Process ID	PL16		
Process Name	Plan communications		
Process Description	Determine the information needs and methods of distribution for all stakeholders.		
Process Purpose Statement	The purpose of *Plan communications* is to determine the information and communication needs of the stakeholders.		
OUTCOMES			
NUMBER	**DESCRIPTION**		
PL16-01	Communication plan likely to achieve its purpose		
PL16-02	Identification and documentation of the approach to communicate most effectively and efficiently with all stakeholders		
BASE PRACTICES (BPs)			
NUMBER	**DESCRIPTION**		**SUPPORTS**
PL16-BP1	Identify stakeholders and interested parties.		PL16-01
PL16-BP2	Perform stakeholder analysis.		PL16-01
PL16-BP3	Identify information needs for stakeholders.		PL16-01
PL16-BP4	Identify mandated information.		PL16-01
PL16-BP5	Determine a suitable plan for meeting information needs.		PL16-01
WORK PRODUCTS			
INPUTS			
NUMBER	**DESCRIPTION**		**SUPPORTS**
PL01-WP1	Project plans		PL16-01
IN02-WP1	Stakeholder register		PL16-01
PL06-WP1	Role descriptions		PL16-01
Change management system	Approved changes		PL16-01
OUTPUTS			
NUMBER	**DESCRIPTION**	**INPUT TO**	**SUPPORTS**
PL16-WP1	Communications plan	IM07, CO11	PL16-01
REFERENCES			
DOCUMENT		**SECTION**	
AA1000		4.1	
COBIT 5 Framework		BAI01	
ISO 21500:2012		4.3.38	
PMBOK 6th Edition		10.1	
PRINCE2		3.2.2	

Implement Domain

1. Direct project work
2. Manage stakeholders
3. Develop project team
4. Treat risks
5. Perform quality assurance
6. Select suppliers
7. Distribute information

Area	Management		
Domain	Implement		
Process ID	IM01		
Process Name	Direct project work		
Process Description	Manage interface between the project sponsor, project manager, and PM team-enabling the work performed by the team to be integrated into further project work or the final project deliverables.		
Process Purpose Statement	The purpose of *Direct project work* is to manage the performance of the work as defined in the project plans to provide the approved project deliverables.		
OUTCOMES			
NUMBER	DESCRIPTION		
IM01-01	Completed work as per schedule		
IM01-02	Overall management of the work		
BASE PRACTICES (BPs)			
NUMBER	DESCRIPTION		SUPPORTS
IM01-BP1	Direct performance of planned project activities		IM01-01
IM01-BP2	Manage technical, administrative, and organizational project interfaces		IM01-01
WORK PRODUCTS			
INPUTS			
NUMBER	DESCRIPTION		SUPPORTS
PL01-WP1	Project plans		IM01-01
Change management system	Approved changes		IM01-01
OUTPUTS			
NUMBER	DESCRIPTION	INPUT TO	SUPPORTS
IM01-WP1	Progress data	CO01, 03, 04, 05, 06, 07, 08	IM01-01 IM01-02
IM01-WP2	Issues log	CO01, CL02	IM01-02
IM01-WP3	Lessons learned	CL02	IM01-02
IM01-WP4	Deliverables	Client process	IM01-01 IM01-02
REFERENCES			
DOCUMENT		SECTION	
COBIT 5 Framework		BAI01	
ISO 21500:2012		4.3.4	
PMBOK 6th Edition		4.3	
PRINCE2		4.1.5	

Area	Management		
Domain	Implement		
Process ID	IM02		
Process Name	Manage stakeholders		
Process Description	Manage all projects by improving communication to stakeholders and involving the business.		
Process Purpose Statement	The purpose of *Manage stakeholders* is to give appropriate understanding and attention to stakeholders' needs and expectations		
OUTCOMES			
NUMBER	DESCRIPTION		
IM02-01	Relevant stakeholders are engaged in projects.		
IM02-02	Continuous communication with stakeholders to understand their needs and expectations.		
BASE PRACTICES (BPs)			
NUMBER	DESCRIPTION		SUPPORTS
IM02-BP1	Identify stakeholders		IM02-01
IM02-BP2	Consider stakeholder interests		IM02-02
IM02-BP3	Plan stakeholder management		IM02-01 IM02-02
IM02-BP4	Manage stakeholder engagement to ensure an active exchange of accurate, consistent, and timely information that reaches all relevant stakeholders		IM02-01 IM02-02
IM02-BP5	Control stakeholder engagement		IM02-02
WORK PRODUCTS			
INPUTS			
NUMBER	DESCRIPTION		SUPPORTS
IN02-WP1	Stakeholder register		IM02-01
PL01-WP1	Project plans		IM02-02
OUTPUTS			
NUMBER	DESCRIPTION	INPUT TO	SUPPORTS
IM02-WP1	Change requests	C004	IM02-02
IM02-WP2	Issue log	Internal	IM02-02

(*Continued*)

REFERENCES	
DOCUMENT	SECTION
AA1000	4.3
COBIT 5 Framework	BAI01
ISO 21500:2012	4.3.10
OPM3	Appendix F
PMBOK 6th Edition	13.3
PRINCE2	3.2

Area	Management	
Domain	Implement	
Process ID	IM03	
Process Name	Develop project team	
Process Description	Assess and document performance of project team members to improve performance and interaction of team members.	
Process Purpose Statement	The purpose of *Develop project team* is to improve the performance and interaction of team members in a continuing manner. This process should enhance team motivation and performance.	
OUTCOMES		
NUMBER	DESCRIPTION	
IM03-01	Ground rules of acceptable behavior	
IM03-02	Improved teamwork	
IM03-03	Advanced people skills and competences.	
IM03-04	Motivated employees.	
IM03-05	Reduced staff turnover rates.	
IM03-06	Improved overall project performance.	
BASE PRACTICES (BPs)		
NUMBER	DESCRIPTION	SUPPORTS
IM03-BP1	Use teamwork	IM03-01 IM03-02 IM03-03 IM03-04
IM03-BP2	Review project plan and deliverables	IM03-04 IM03-05 IM03-06
IM03-BP3	Review staff assignments	IM03-04 IM03-05 IM03-06
IM03-BP4	Assess performance	IM03-06
WORK PRODUCTS		
INPUTS		
NUMBER	DESCRIPTION	SUPPORTS
IN03-WP1	Staff assignments	IM03-01
Human resource management systems	Resource availability	IM03-01
PL06-WP1	Role descriptions	IM03-01

(*Continued*)

OUTPUTS			
NUMBER	DESCRIPTION	INPUT TO	SUPPORTS
IM03-WP1	Team performance	Performance measurement system	IM03-01
IM03-WP2	Team appraisal	Performance measurement system	IM03-01

REFERENCES	
DOCUMENT	SECTION
COBIT 5 Framework	BAI01
ISO 21500:2012	4.3.18
PMBOK 6th Edition	9.4

Area	Management		
Domain	Implement		
Process ID	IM04		
Process Name	Treat risks		
Process Description	Continually respond to risks by treatment and ensure risks are within risk tolerance allocated to the project by the project steering committee/project Governing Body.		
Process Purpose Statement	The purpose of *Treat risks* is to develop options and determine actions to enhance opportunities and reduce threats to project objectives.		
OUTCOMES			
NUMBER	DESCRIPTION		
IM04-01	Continually treat risk to ensure risk is within risk tolerance limits.		
IM04-02	All significant risks are managed.		
IM04-03	Risk activities are carried out effectively.		
BASE PRACTICES (BPs)			
NUMBER	DESCRIPTION		SUPPORTS
IM04-BP1	Reduce risk to an acceptable level.		IM04-01 IM04-02 IM04-03
IM04-BP2	Eliminate or minimize project risk through a systemic process.		IM04-02
IM04-BP3	Respond in a timely manner with effective countermeasures to limit loss from projects.		IM04-01 IM04-02 IM04-03
WORK PRODUCTS			
INPUTS			
NUMBER	DESCRIPTION		SUPPORTS
PL12-WP1	Risk register		IM04-01
PL01-WP1	Project plans		IM04-01
OUTPUTS			
NUMBER	DESCRIPTION	INPUT TO	SUPPORTS
IM04-WP1	Risk responses	CO08	IM04-01
IM04-WP2	Change requests	Change management system	IM04-01
REFERENCES			
DOCUMENT		SECTION	
COBIT 5 Framework		APO12 BAI01	
ISO 21500:2012		4.3.30	
PMBOK 6th Edition		11.5	
PRINCE2		3.5	

Area	Management	
Domain	Implement	
Process ID	IM05	
Process Name	Perform quality assurance	
Process Description	Monitor projects for the use of approved practices and standards in continuous improvement and efficiency efforts.	
Process Purpose Statement	The purpose of *Perform quality assurance* is to review the deliverables and the project. It includes all processes, tools, procedures, techniques and resources necessary to meet quality requirements. This process includes the following: • Ensure objectives and relevant standards to be achieved are communicated, understood, accepted, and adhered to by the appropriate project organization members. • Execute the quality plan as the project progresses. • Ensure that the established tools, procedures, techniques, and resources are being used.	
OUTCOMES		
NUMBER	DESCRIPTION	
IM05-01	Project delivery results are predictable.	
IM05-02	Quality requirements are implemented in all processes.	
BASE PRACTICES (BPs)		
NUMBER	DESCRIPTION	SUPPORTS
IM05-BP1	Improve quality to achieve customer satisfaction.	IM05-01
IM05-BP2	Project initiation process improvement	IM05-01
IM05-BP3	Project plan development process improvement	IM05-01
IM05-BP4	Project scope planning process improvement	IM05-01
IM05-BP5	Project scope definition process improvement	IM05-01
IM05-BP6	Project activity definition process improvement	IM05-01
IM05-BP7	Project activity sequencing process improvement	IM05-01
IM05-BP8	Project activity duration estimating process improvement	IM05-01
IM05-BP9	Project schedule development process improvement	IM05-01
IM05-BP10	Project resource planning process improvement	IM05-01
IM05-BP11	Project cost estimating process improvement	IM05-01
IM05-BP12	Project cost budgeting process improvement	IM05-01
IM05-BP13	Project risk management planning process improvement	IM05-01
IM05-BP14	Project quality planning process improvement	IM05-01
IM05-BP15	Project organizational planning process improvement	IM05-01
IM05-BP16	Project staff acquisition process improvement	IM05-01
IM05-BP17	Project communications planning process improvement	IM05-01
IM05-BP18	Project risk identification process improvement	IM05-01
IM05-BP19	Project qualitative risk analysis process improvement	IM05-01

(*Continued*)

NUMBER	DESCRIPTION		SUPPORTS
IM05-BP20	Project quantitative risk analysis process improvement		IM05-01
IM05-BP21	Project risk response planning process improvement		IM05-01
IM05-BP22	Project procurement planning process improvement		IM05-01
IM05-BP23	Project solicitation planning process improvement		IM05-01
IM05-BP24	Project plan execution process improvement		IM05-01
IM05-BP25	Project quality assurance process improvement		IM05-01
IM05-BP26	Project team development process improvement		IM05-01
IM05-BP27	Project information distribution process improvement		IM05-01
IM05-BP28	Project solicitation process improvement		IM05-01
IM05-BP29	Project source selection process improvement		IM05-01
IM05-BP30	Project contract administration process improvement		IM05-01
IM05-BP31	Project performance reporting process improvement		IM05-01
IM05-BP32	Project integrated change control process improvement		IM05-01
IM05-BP33	Project scope verification process improvement		IM05-01
IM05-BP34	Project scope change control process improvement		IM05-01
IM05-BP35	Project schedule control process improvement		IM05-01
IM05-BP36	Project cost control process improvement		IM05-01
IM05-BP37	Project quality control process improvement		IM05-01
IM05-BP38	Project risk monitoring and control process improvement		IM05-01
IM05-BP39	Project contract closeout process improvement		IM05-01
IM05-BP40	Project administrative closure process improvement		IM05-01
WORK PRODUCTS			
INPUTS			
NUMBER	DESCRIPTION		SUPPORTS
PL14-WP1	Quality plan		IM05-01
OUTPUTS			
NUMBER	DESCRIPTION	INPUT TO	SUPPORTS
IM05-WP1	Change requests	Change management system	IM05-01
REFERENCES			
DOCUMENT		SECTION	
COBIT 5 Framework		APO11 BAI01	
ISO 21500:2012		4.3.33	
PMBOK 6th Edition		8.2	
PRINCE2		3.3	

Area	Management	
Domain	Implement	
Process ID	IM06	
Process Name	Select suppliers	
Process Description	Develop process for selecting suppliers based on criteria and use the process and criteria to select suppliers in a consistent manner that complies with any legal, regulatory or contractual requirements.	
Process Purpose Statement	The purpose of *Select suppliers* is: • To ensure that information is obtained from suppliers so that there is consistent evaluation of proposals against stated requirements. • To review and examine all the submitted information. • To select the suppliers.	
OUTCOMES		
NUMBER	DESCRIPTION	
IM06-01	Unambiguous RFIs, RFPs or RFQs.	
IM06-02	Selected suppliers meeting stated requirements and providing the greatest benefit	
IM06-03	Alignment of internal and external stakeholder expectations through established agreements	
BASE PRACTICES (BPs)		
NUMBER	DESCRIPTION	SUPPORTS
IM06-BP1	Develop methods for selecting suppliers.	IM06-01
IM06-BP2	Develop criteria for selecting suppliers.	IM06-01
IM06-BP3	Evaluate supplier submissions based on evaluation criteria.	IM06-01
IM06-BP4	Select suppliers.	IM06-02 IM06-03
IM06-BP5	Negotiate with suppliers.	IM06-02 IM06-03
IM06-BP6	Award contract based on final agreement and conditions.	IM06-03
WORK PRODUCTS		
INPUTS		
NUMBER	DESCRIPTION	SUPPORTS
PL15-WP1	Procurement plan	IM06-01
Vendor management system	Preferred supplier list	IM06-02
Vendor management system	Supplier's tenders	IM06-02
PL15-WP3	Make-or-buy decision list	IM06-01

(*Continued*)

OUTPUTS			
NUMBER	DESCRIPTION	INPUT TO	SUPPORTS
IM06-WP1	Request for information, proposal, bid, offer, or quotation	Procurement	IM06-01
IM06-WP2	Contracts or purchase orders	CO10	IM06-01
IM06-WP3	Selected suppliers list	Vendor management system	IM06-01

REFERENCES	
DOCUMENT	SECTION
COBIT 5 Framework	APO10 BAI01
ISO 21500:2012	4.3.36
PMBOK 6th Edition	12.2.3.1

Area	Management		
Domain	Implement		
Process ID	IM07		
Process Name	Distribute information		
Process Description	Distribute information based on the communication plan and amend or provide organizational policies, procedures, and other information as required.		
Process Purpose Statement	The purpose of *Distribute information* is to make required information available to project stakeholders, as defined by the communications plan, and to respond to unexpected, specific requests for information.		
OUTCOMES			
NUMBER	DESCRIPTION		
IM07-01	Effective and efficient flow between project stakeholders		
BASE PRACTICES (BPs)			
NUMBER	DESCRIPTION		SUPPORTS
IM07-BP1	Generate appropriate information.		IM07-01
IM07-BP2	Choose media for distribution.		IM07-01
IM07-BP3	Develop feedback model.		IM07-01
IM07-BP4	Communicate information to stakeholders.		IM07-01
IM07-BP5	Hold productive meetings with stakeholders.		IM07-01
WORK PRODUCTS			
INPUTS			
NUMBER	DESCRIPTION		SUPPORTS
PL16-WP1	Communications plan		IM07-01
CO01-WP2	Progress reports		IM07-01
Governing Body	Unexpected requests		IM07-01
OUTPUTS			
NUMBER	DESCRIPTION	INPUT TO	SUPPORTS
IM07-WP1	Distributed information	CO11	IM07-01
REFERENCES			
DOCUMENT		SECTION	
AA1000		4.3	
COBIT 5 Framework		BAI01	
ISO 21500:2012		4.3.39	
PMBOK 6th Edition		10.2	

Control Domain

1. Control project work
2. Control changes
3. Control scope
4. Control resources
5. Manage project team
6. Control schedule
7. Control costs
8. Control risks
9. Perform quality control
10. Administer procurements
11. Manage communications

Area	Management
Domain	Control
Process ID	C001
Process Name	Control project work
Process Description	Manage all project work in a coordinated way.
Process Purpose Statement	The purpose of *Control project work* is to complete project activities in an integrated manner in accordance with the project plans
OUTCOMES	
NUMBER	DESCRIPTION
C001-01	Project activities are executed according to the plans.
C001-02	There are sufficient project resources to perform activities according to the plans.
C001-03	Project expected benefits are achieved and accepted.

(*Continued*)

BASE PRACTICES (BPs)		
NUMBER	DESCRIPTION	SUPPORTS
C001-BP1	Manage using project processes.	C001-01
C001-BP2	Project initiation process measurement	C001-01
C001-BP3	Project plan development process measurement	C001-01
C001-BP4	Project activity definition process measurement.	C001-01
C001-BP5	Project activity sequencing process measurement	C001-01
C001-BP6	Project activity duration estimating process measurement	C001-01
C001-BP7	Benchmark project performance against industry standards.	C001-01
C001-BP8	Project initiation process control.	C001-01
C001-BP9	Project plan development process control.	C001-01
C001-BP10	Project organizational planning process control.	C001-01
C001-BP11	Project plan execution process control.	C001-01
C001-BP12	Project administrative closure process control.	C001-01

WORK PRODUCTS		
INPUTS		
NUMBER	DESCRIPTION	SUPPORTS
PL01-WP1	Project plan	C001-01
IM01-WP1	Progress data	C001-01
Quality management system	Quality control measurements	C001-01
PL12-WP1	Risk register	C001-01
IM01-WP2	Issues log	C001-01

OUTPUTS			
NUMBER	DESCRIPTION	INPUT TO	SUPPORTS
C001-WP1	Change requests	C007-03	C001-01
C001-WP2	Progress reports	IM07, CL01, CL02	C001-01
C001-WP3	Project completion reports	CL01	C001-01

REFERENCES	
DOCUMENT	SECTION
COBIT 5 Framework	BAI01
ISO 21500:2012	4.3.5
OPM3	Appendix F
PMBOK 6th Edition	4.6

Area	Management		
Domain	Control		
Process ID	C002		
Process Name	Control changes		
Process Description	Manage all changes in a controlled manner, including scope, time, and cost changes related to the project. This includes impact assessment, prioritization and authorization, tracking, reporting, closure, and documentation.		
Process Purpose Statement	The purpose of *Control changes* is to control changes to the project and deliverables and to formalize acceptance, deferral, or rejection of these changes before subsequent implementation.		
OUTCOMES			
NUMBER	DESCRIPTION		
C002-01	Authorized changes are made in a timely manner.		
C002-02	Impact assessments reveal the impact on the project.		
C002-03	Key stakeholders are kept informed of all aspects of the change.		
BASE PRACTICES (BPs)			
NUMBER	DESCRIPTION		SUPPORTS
C002-BP1	Evaluate all changes to determine the impact on the project.		C002-01
C002-BP2	Project integrated change control process control		C002-01
C002-BP2	Project scope change control process control		C002-02
C002-BP3	Track and report change status.		C002-03
C002-BP4	Close and document all changes.		C002-01
WORK PRODUCTS			
INPUTS			
NUMBER	DESCRIPTION		SUPPORTS
PL01-WP1	Project plans		C002-01
Change management system	Change requests		C002-01
OUTPUTS			
NUMBER	DESCRIPTION	INPUT TO	SUPPORTS
C002-WP1	Impact assessments	Internal	C002-01 C002-BP1
C002-WP2	Approved changes	Change management system	C002-01 C002-BP2 C002-BP3

(*Continued*)

NUMBER	DESCRIPTION	INPUT TO	SUPPORTS
C002-WP3	Change register	Internal	C002-01 C002-BP3
C002-WP4	Change documentation	Internal	C002-03 C002-BP2

REFERENCES	
DOCUMENT	SECTION
COBIT 5 Framework	BAI01, BAI06
ISO 21500:2012	4.3.6
PMBOK 6th Edition	4.6

Area	Management		
Domain	Control		
Process ID	CO03		
Process Name	Control scope		
Process Description	Manage scope and limit scope creep or unplanned scope changes.		
Process Purpose Statement	The purpose of *Control scope* is to maximize positive and minimize negative project impacts created by scope changes.		
OUTCOMES			
NUMBER	DESCRIPTION		
CO03-01	Project scope that adheres to the baseline scope.		
CO03-02	Approved and controlled changes to project scope.		
Base Practices (BPs)			
NUMBER	DESCRIPTION		SUPPORTS
CO03-BP1	Project scope planning process measurement		CO03-01 CO03-02
CO03-BP2	Project scope definition process measurement		CO03-01 CO03-02
CO03-BP3	Project scope planning process control		CO03-01 CO03-02
CO03-BP4	Project scope verification process control		CO03-01 CO03-02
WORK PRODUCTS			
INPUTS			
NUMBER	DESCRIPTION		SUPPORTS
IM01-WP1	Progress data		CO03-01
PL02-WP1	Scope statement		CO03-01
PL03-WP1	Work breakdown structure		CO03-01
PL04-WP1	Activity list		CO03-01
OUTPUTS			
NUMBER	DESCRIPTION	INPUT TO	SUPPORTS
CO03-WP1	Change requests	Change management system	CO03-01
REFERENCES			
DOCUMENT		SECTION	
COBIT 5 Framework		BAI01	
ISO 21500:2012		4.3.14	
OPM3		Appendix F	
PMBOK 6th Edition		5.6	

Area	Management		
Domain	Control		
Process ID	C004		
Process Name	Control resources		
Process Description	Manage project resources to minimize the impact of conflicts of availability or unavailability by rescheduling activities or finding alternative resources.		
Process Purpose Statement	The purpose of *Control resources* is to ensure that the resources required to undertake the project work are available and assigned in the manner necessary to meet the project requirements.		
OUTCOMES			
NUMBER	DESCRIPTION		
C004-01	Resource availability as per schedule.		
BASE PRACTICES (BPs)			
NUMBER	DESCRIPTION		SUPPORTS
C004-BP1	Use formal performance assessment.		C004-01
C004-BP2	Record project resource requirements.		C004-01
C004-BP3	Project resource planning process measurement		C004-01
C004-BP4	Project activity definition process control		C004-01
C004-BP5	Project activity sequencing process control		C004-01
C004-BP6	Project activity duration process control		C004-01
C004-BP7	Project resource planning process control		C004-01
C004-BP8	Project communications planning process control		C004-01
WORK PRODUCTS			
INPUTS			
NUMBER	DESCRIPTION		SUPPORTS
PL01-WP1	Project plans		C004-01
Human resource management system	Staff assignments		C004-01
Human resource management system	Resource availability		C004-01
IM01-WP1	Progress data		C004-01
PL05-WP1	Resource requirements		C004-01
OUTPUTS			
NUMBER	DESCRIPTION	INPUT TO	SUPPORTS
IM02-WP1	Change requests	Change management system	C004-01
C004-WP4	Corrective actions	Quality management system	C004-01

(*Continued*)

REFERENCES	
DOCUMENT	SECTION
COBIT 5 Framework	BAI01
ISO 21500:2012	4.3.19
OPM3	Appendix F
PMBOK 6th Edition	9.6

Area	Management	
Domain	Control	
Process ID	C005	
Process Name	Manage project team	
Process Description	Manage the project team and resource requirements and raise issues for performance appraisal or lessons learned.	
Process Purpose Statement	The purpose of *Manage project team* is to optimize team performance, provide feedback, resolve issues, encourage communication and coordinate changes to achieve project success.	
OUTCOMES		
NUMBER	DESCRIPTION	
C005-01	Successful project	
C005-02	Managed conflicts and issues	
C005-03	Optimized project performance	
C005-04	Appraised team member performance	
BASE PRACTICES (BPs)		
NUMBER	DESCRIPTION	SUPPORTS
C005-BP1	Project staff acquisition process control	C005-01 C005-03
C005-BP2	Project team development process control	C005-01 C005-04
C005-BP3	Supportive team environment	C005-02
C005-BP4	Build trust.	C005-01 C005-02 C005-03
C005-BP5	Provide technical administrative support.	C005-01 C005-03
WORK PRODUCTS		
INPUTS		
NUMBER	DESCRIPTION	SUPPORTS
PL01-WP1	Project plans	C005-01
PL06-WP2	Project organization chart	C005-01 C005-02
PL06-WP1	Role descriptions	C005-01 C005-02
IM01-WP1	Progress data	C005-01 C005-03
Resource planning system	Resource management plan	C005-03

(*Continued*)

OUTPUTS			
NUMBER	DESCRIPTION	INPUT TO	SUPPORTS
C005-WP1	Staff performance	Performance measurement system	C005-01
C005-WP2	Staff appraisals	Performance measurement system	C005-04
C004-WP3	Change requests	Change management system	C005-01 C005-03
C004-WP4	Corrective actions	C003-01	C005-01 C005-03

REFERENCES	
DOCUMENT	SECTION
COBIT 5 Framework	BAI01
ISO 21500:2012	4.3.20
OPM3	Appendix F
PMBOK 6th Edition	9.4

Area	Management		
Domain	Control		
Process ID	C006		
Process Name	Control schedule		
Process Description	Determine the current schedule status, forecast completion dates, and make adjustments to handle any project variances or avoid adverse impacts.		
Process Purpose Statement	The purpose of *Control schedule* is to monitor schedule variances and to take appropriate actions.		
OUTCOMES			
NUMBER	DESCRIPTION		
C006-01	Project or project phase delivered on time		
C006-02	Deviations recognized and corrective or preventive action taken		
C006-03	Minimized project risk		
BASE PRACTICES (BPs)			
NUMBER	DESCRIPTION		SUPPORTS
C006-BP1	Project schedule development process measurement		C006-01
C006-BP2	Project schedule development process control		C006-01
C006-BP3	Project schedule control process control		C006-01
C006-BP4	Determine status of the project.		C006-01
C006-BP5	Take corrective action.		C006-02 C006-03
WORK PRODUCTS			
INPUTS			
NUMBER	DESCRIPTION		SUPPORTS
PL09-WP1	Schedule		C006-01 C006-02
IM01-WP1	Progress data		C006-01 C006-02
PL01-WP1	Project plans		C006-01 C006-02
OUTPUTS			
NUMBER	DESCRIPTION	INPUT TO	SUPPORTS
C006-WP1	Change requests	Change management system	C006-01
C006-WP2	Corrective actions	Quality management system	C006-02 C006-03

(*Continued*)

REFERENCES	
DOCUMENT	SECTION
COBIT 5 Framework	BAI01
ISO 21500:2012	4.3.24
PMBOK 6th Edition	6.6

Area	Management	
Domain	Control	
Process ID	C007	
Process Name	Control costs	
Process Description	Determine current costs and any variances, forecast projected costs at completion, and take any appropriate preventive or corrective action to avoid cost impact.	
Process Purpose Statement	The purpose of *Control costs* is to monitor cost variances and to take appropriate actions.	
OUTCOMES		
NUMBER	DESCRIPTION	
C007-01	Project delivered as per agreed budget	
C007-02	Recognition of cost variance	
C007-03	Corrective action taken to minimize risk	
BASE PRACTICES		
NUMBER	DESCRIPTION	SUPPORTS
C007-BP1	Project cost estimating process control	C007-01
C007-BP2	Project cost budgeting process control	C007-01
C007-BP3	Project cost control process control	C007-02
C007-BP4	Act on change requests in a timely manner.	C007-03
C007-BP5	Manage changes.	C007-03
C007-BP6	Ensure cost expenditures do not exceed authorized funding or contingency reserves.	C007-02
C007-BP7	Monitor cost performance.	C007-01 C007-02
WORK PRODUCTS		
INPUTS		
NUMBER	DESCRIPTION	SUPPORTS
IM01-WP1	Progress data	C007-01 C007-02
PL01-WP1	Project plans	C007-01
PL11-WP1	Budget	C007-01 C007-02

(*Continued*)

OUTPUTS			
NUMBER	DESCRIPTION	INPUT TO	SUPPORTS
C007-WP1	Actual costs	Financial management system	C007-01 C007-02
C007-WP2	Forecasted costs	Budgeting system	C007-01 C007-02 C007-03
C007-WP3	Change requests	Change management system	C007-03
C007-WP4	Corrective actions	Quality management system	C007-02 C007-03

REFERENCES	
DOCUMENT	SECTION
COBIT 5 Framework	APO06 BAI01
ISO 21500:2012	4.3.27
PMBOK 6th Edition	7.4

Area	Management		
Domain	Control		
Process ID	CO08		
Process Name	Control risks		
Process Description	Ensure that risk responses are implemented efficiently and assess the effectiveness of the response.		
Process Purpose Statement	The purpose of *Control risks* is to minimize disruption to the project by determining whether the risk responses are executed and whether they have the desired effect.		
OUTCOMES			
NUMBER	DESCRIPTION		
CO08-01	Risk responses are carried out efficiently.		
CO08-02	Risk responses are effective and reduce risk and enhance opportunities.		
CO08-03	Improved efficiency of the risk approach throughout the project.		
BASE PRACTICES (BPs)			
NUMBER	DESCRIPTION		SUPPORTS
CO08-BP1	Project risk management planning process control		CO08-01
CO08-BP2	Project risk identification process control		CO08-01
CO08-BP3	Project qualitative risk analysis process control		CO08-01
CO08-BP4	Project quantitative risk analysis process control		CO08-01
CO08-BP5	Project risk response planning process control		CO08-01
CO08-BP6	Project risk monitoring and control process control		CO08-01
WORK PRODUCTS			
INPUTS			
NUMBER	DESCRIPTION		SUPPORTS
PL12-WP1	Risk register		CO08-01
IM01-WP1	Progress data		CO08-01
PL01-WP1	Project plans		CO08-01
IM04-WP1	Risk responses		CO08-01
OUTPUTS			
NUMBER	DESCRIPTION	INPUT TO	SUPPORTS
CO08-WP1	Change requests	Change management system	CO08-01
CO08-WP2	Corrective actions	Quality management system	CO08-01

(*Continued*)

REFERENCES	
DOCUMENT	SECTION
COBIT 5 Framework	APO12 BAI01
ISO 21500:2012	4.3.31
OPM3	Appendix F
PMBOK 6th Edition	11.6

Area	Management
Domain	Control
Process ID	C009
Process Name	Perform quality control
Process Description	Ensure project objectives, quality requirements and standards are met efficiently and effectively.
Process Purpose Statement	The purpose of *Perform quality control* is to determine whether the project's established objectives, quality requirements and standards are being met and to identify causes of, and ways to eliminate, unsatisfactory performance.

OUTCOMES

NUMBER	DESCRIPTION
C009-01	Quality requirements are performed in all activities
C009-02	Improvement of all processes

BASE PRACTICES (BPs)

NUMBER	DESCRIPTION	SUPPORTS
C009-BP1	Measure consistently	C009-01
C009-BP2	Measure accurately	C009-01
C009-BP3	Project quality planning process control	C009-01
C009-BP4	Project quality assurance process control	C009-01
C009-BP5	Project performance reporting process	C009-01
C009-BP6	Project quality control process control	C009-01
C009-BP7	Perform benchmarking to improve performance	C009-01

WORK PRODUCTS

INPUTS

NUMBER	DESCRIPTION	SUPPORTS
IM01-WP1	Progress data	C009-01
PL01-WP1	Customer requirements for project quality	C009-01
PL01-WP1	Acceptance criteria	C009-01
PL14-WP1	Quality plan	C009-01

OUTPUTS

NUMBER	DESCRIPTION	INPUT TO	SUPPORTS
C009-WP1	Quality control measurements	Quality management system	C009-01
C009-WP2	Verified deliverables	Business process	C009-01
C009-WP3	Inspection reports	C010	C009-01
C009-WP4	Change requests	Change management system	C009-01
C009-WP5	Corrective actions	Quality management system	C009-01

(*Continued*)

REFERENCES	
DOCUMENT	SECTION
COBIT 5 Framework	APO11 BAI01
ISO 21500:2012	4.3.34
OPM3	Appendix F
PMBOK 6th Edition	8.2

Area	Management		
Domain	Control		
Process ID	CO10		
Process Name	Administer procurements		
Process Description	Monitoring and reviewing suppliers' performance and receipt of progress reports as per the contract and taking appropriate action when a supplier is non-compliant.		
Process Purpose Statement	The purpose of *Administer procurements* is to manage the relationship between the buyer and the suppliers.		
OUTCOMES			
NUMBER	**DESCRIPTION**		
CO10-01	Seller's and buyer's performance meet procurement requirements according to terms of the contract or agreement		
BASE PRACTICES (BPs)			
NUMBER	**DESCRIPTION**		**SUPPORTS**
CO10-BP1	Project procurement planning process control		CO10-01
CO10-BP2	Project solicitation planning process control		CO10-01
CO10-BP3	Project solicitation process control		CO10-01
CO10-BP4	Project source selection process control		CO10-01
CO10-BP5	Project contract administration process control		CO10-01
CO10-BP6	Project contract closeout process control		CO10-01
WORK PRODUCTS			
INPUTS			
NUMBER	**DESCRIPTION**		**SUPPORTS**
IM06-WP2	Contracts or purchase orders		CO10-01
PL01-WP1	Project plans		CO10-01
Change management system	Approved changes		CO10-01
CO09-WP3	Inspection reports		CO10-01
OUTPUTS			
NUMBER	**DESCRIPTION**	**INPUT TO**	**SUPPORTS**
CO10-WP1	Change requests	Change management system	CO10-01
CO10-WP2	Corrective actions	Internal	CO10-01

(*Continued*)

REFERENCES	
DOCUMENT	SECTION
COBIT 5 Framework	BAI01
ISO 21500:2012	4.3.37
OPM3	Appendix F
PMBOK 6th Edition	12.3

Area	Management		
Domain	Control		
Process ID	CO11		
Process Name	Manage communications		
Process Description	Focus on increasing the understanding and cooperation amongst stakeholders; provide timely, accurate, and unbiased information; and resolve communication issues to minimize risk to the project.		
Process Purpose Statement	The purpose of *Manage communications* is to ensure that the communication needs of the project stakeholders are satisfied and to resolve communication issues when they arise.		
OUTCOMES			
NUMBER	DESCRIPTION		
CO11-01	Optimal communications flow between stakeholders		
BASE PRACTICES (BPs)			
NUMBER	DESCRIPTION		SUPPORTS
CO11-BP1	Communicate the organization's direction.		CO11-01
CO11-BP2	Project information distribution process control		CO11-01
WORK PRODUCTS			
INPUTS			
NUMBER	DESCRIPTION		SUPPORTS
PL16-WP1	Communications plan		CO11-01
IM07-WP1	Distributed information		CO11-01
OUTPUTS			
NUMBER	DESCRIPTION	INPUT TO	SUPPORTS
CO11-WP1	Accurate and timely information	Stakeholders	CO11-01
CO11-WP2	Corrective actions	Quality management system	CO11-01
REFERENCES			
DOCUMENT		SECTION	
AA1000		4.3.6	
COBIT 5 Framework		BAI01	
ISO 21500:2012		4.3.40	
OPM3		Appendix F	
PMBOK 6th Edition		10.3	

Close Domain

1. Close project phase or project
2. Collect lessons learned

Area	Management	
Domain	Close	
Process ID	CL01	
Process Name	Close project phase or project	
Process Description	Ensure the phase or project is closed in a structured manner and communicate outstanding activities.	
Process Purpose Statement	The purpose of *Close project phase or project* is to confirm the completion of all project processes and activities in order to close a project phase or a project.	
OUTCOMES		
NUMBER	DESCRIPTION	
CL01-01	The project is closed, and expected benefits are achieved and accepted.	
BASE PRACTICES (BPs)		
NUMBER	DESCRIPTION	SUPPORTS
CL01-BP1	Define and apply steps for project or phase closure.	CL01-01
CL01-BP2	Require the project stakeholders to ascertain whether the project, release, sprint, or iteration delivered the planned results and value.	CL01-01
CL01-BP3	Plan and execute post-implementation reviews to determine whether the project met its objectives.	CL01-01
CL01-BP4	Identify and communicate any outstanding activities required to achieve the planned results or benefits of the project.	CL01-01
CL01-BP5	Obtain stakeholder acceptance of project products or deliverables and transfer ownership.	CL01-01
WORK PRODUCTS		
INPUTS		
NUMBER	DESCRIPTION	SUPPORTS
ES01-WPX	Requirements for stage-gate reviews	CL01-01 CL01-BP1
CO01-WP2	Progress reports	CL01-01 CL01-BP2
Contract Management System	Contract documentation	CL01-01 CL01-BP1
CO01-WP3	Project completion reports	CL01-01 CL01-BP2

(*Continued*)

OUTPUTS			
NUMBER	DESCRIPTION	INPUT TO	SUPPORTS
CL01-WP1	Completed procurements	Vendor management system	CL01-01 CL01-BP1
CL01-WP2	Project or phase closure report	Project stakeholders	CL01-01 CL01-BP2
CL01-WP3	Released resources	Business units	CL01-01 CL01-BP2
CL01-WP4	Final product, service, or result transition	Stakeholders	CL01-BP5

REFERENCES	
DOCUMENT	SECTION
COBIT 5 Framework	BAI01
ISO 21500:2012	4.3.7
PMBOK 6th Edition	4.7
PRINCE2	4.6

Area	Management		
Domain	Close		
Process ID	CL02		
Process Name	Collect lessons learned		
Process Description	Identify, collect, and document lessons learned for use on future projects, phases, sprints, releases, or iteration.		
Process Purpose Statement	The purpose of *Collect lessons learned* is to evaluate the project and collect experiences that would benefit current and future projects.		
OUTCOMES			
NUMBER	DESCRIPTION		
CL02-01	Lessons from the project or project component		
BASE PRACTICES (BPs)			
NUMBER	DESCRIPTION		SUPPORTS
CL02-BP1	Quantify lessons learned.		CL02-01
CL02-BP2	Capture and share lessons learned.		CL02-01
CL02-BP3	Apply lessons learned.		CL02-01
WORK PRODUCTS			
INPUTS			
NUMBER	DESCRIPTION		SUPPORTS
PL01-WP1	Project plans		CL02-01
CO01-WP2	Progress reports		CL02-01
Change management system	Approved changes		CL02-01
IM01-WP3	Lessons learned		CL02-01
IM01-WP2	Issues log		CL02-01
PL12-WP1	Risk register		CL02-01
OUTPUTS			
NUMBER	DESCRIPTION	INPUT TO	SUPPORTS
CL02-WP1	Lessons learned document	Knowledge management system	CL02-01
REFERENCES			
DOCUMENT		SECTION	
COBIT 5 Framework		BAI01, BAI08	
ISO 21500:2012		4.3.8	
OPM3		Appendix F	
PMBOK 6th Edition		8.3, 10.2, 12.3, 13.3, 13.4	

4
THE CAPABILITY DIMENSION

The process capability indicators represent the way to achieve the capability of the process attribute (PA). This really means that any evidence of process indicators supports the assessor's judgement of the degree to which the process meets its purpose.

Technically there are six levels: Level 0 to Level 5. However, there are no process capability indicators for Level 0 as this level represents a non-implemented process or a process that fails to meet its outcomes. Therefore, the remainder of this section describes the process capability indicators for the nine process attributes included in the capability dimension for Levels 1 through 5. The levels and process indicators are shown in Figure 4.1.

The process reference model for the process dimension is used for this assessment. For all the attributes that follow, you will see a chart with three columns. The first column—Result of Full Achievement of the Attribute—is the same for all processes as are the Best Practices and General Work Practices. You should now know that the difference between doing this assessment and an assessment of any other process lies with the process reference model used to assess Level 1. An assessment is an assessment. Once you gain accreditation for assessing one framework, all you need do is have subject matter expertise in another domain. The skills are transferable. Now, it is time to look at the process indicators before going on to perform an assessment in Part II.

Level 0: Incomplete Process

The process is not implemented at all or fails to achieve its purpose. There is little or no objective evidence to show any achievement of the process purpose being accomplished in an organized way.

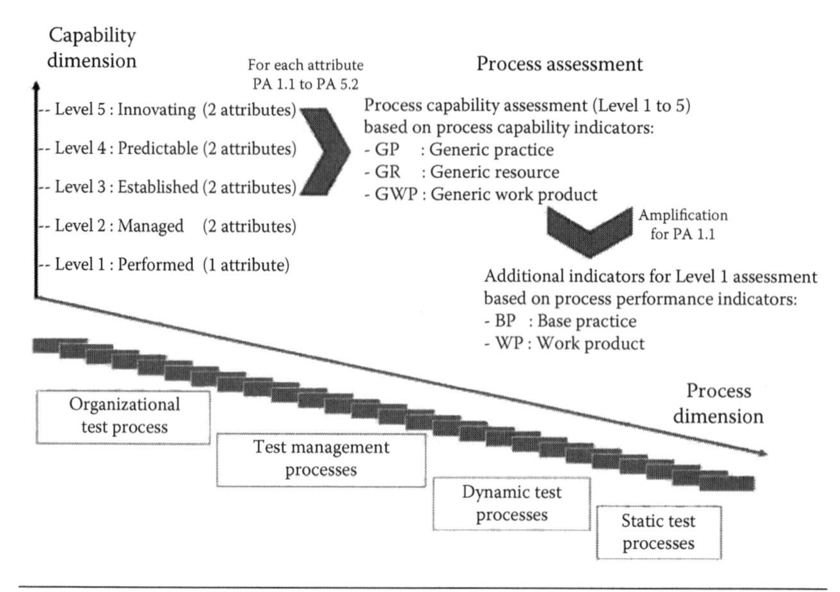

Figure 4.1 Capability indicators and process levels. (Permission to use extracts from ISO 33063:2015 was provided by the Standards Council of Canada [SCC]. No further reproduction is permitted without prior written approval from SCC.)

Level 1: Performed Process

The implemented process achieves its process purpose. To achieve this level, the organization must meet the following process attribute.

PA 1.1 Process Performance

This process attribute is a measure of process purpose achievement. Full achievement of this attribute results in the process achieving its defined outcomes (see Table 4.1).

Table 4.1 Process Attribute 1.1 Process Performance

RESULT OF FULL ACHIEVEMENT OF THE ATTRIBUTE	BEST PRACTICES (BPs)	WORK PRODUCTS (WPs)
The process achieves its outcomes.	*BP 1.1.1 Achieve the process outcomes.* There is evidence that the intent of the base practice is being performed.	WPs are produced that provide evidence of expected outcomes as outlined in the process dimension.

Level 2: Managed Process

The process referenced in Level 1: *Performed Process* can be seen as managed: that is, it is planned, monitored, adjusted, with work products that are properly established, controlled, and maintained. To achieve Level 2, the organization must meet the following PAs.

PA 2.1 Performance Management

This process attribute is a measure of the extent to which the performance of the process is managed. Full achievement of this attribute results in the process achieving the following attributes (see Table 4.2).

Table 4.2 Process Attribute 2.1 Performance Management

RESULT OF FULL ACHIEVEMENT OF THE ATTRIBUTE	GENERIC PRACTICES (GPs)	GENERIC WORK PRODUCTS (GWPs)
a. Process performance objectives are identified.	*GP 2.1.1 Identify the objectives* along with any assumptions and constraints, are defined and communicated.	*GWP 1.0 Process documentation* should outline the process scope. *GWP 2.0 Process plan* should provide details of the process performance objectives.
b. Performance is planned.	*GP 2.1.2 Plan the performance of the process to fulfill the identified objectives.* Basic measures of process performance linked to business objectives are established. They include key milestones, required activities, estimates and schedules.	*GWP 2.0 Process Plan* should provide details of the process performance objectives. *GWP 9.0 Process performance records* should provide details of the outcomes.
c. Performance is monitored.	*GP 2.1.3 Monitor the performance of the process to fulfill the identified objectives.* Basic measures of process performance linked to business objectives are established and monitored. They include key milestones, required activities, estimates, and schedules.	*GWP 2.0 Process Plan* should provide details of the process performance objectives. *GWP 9.0 Process performance records* should provide details of the outcomes.

(*Continued*)

Table 4.2 (*Continued*) Process Attribute 2.1 Performance Management

RESULT OF FULL ACHIEVEMENT OF THE ATTRIBUTE	GENERIC PRACTICES (GPs)	GENERIC WORK PRODUCTS (GWPs)
d. Performance is adjusted to meet plans.	*GP 2.1.4 Adjust the performance of the process.* Action is taken when planned performance is not achieved. Actions include identification of process performance issues and adjustment of plans and schedules as appropriate.	*GWP 4.0 Quality record* should provide details of action taken when performance is not achieved.
e. Define, assign and communicate performance responsibilities and authorities.	*GP 2.1.5 Define responsibilities and authorities for performing the process.* The key responsibilities and authorities for performing the key activities of the process are defined, assigned, and communicated. The need for process performance experience, knowledge, and skills is defined.	*GWP 1.0 Process documentation* should provide details of the process owner and who is responsible, accountable, consulted, and informed (RACI). *GWP 2.0 Process plan* should include details of the process communication plan as well as process performance experience, skills, and requirements.
f. Personnel performing the process are properly prepared.	*GP 2.1.6 Assign personnel and train them to perform the process according to plan.* Personnel performing processes accept responsibility and have the necessary skills and experience to carry out their responsibilities.	*GWP 2.0 Process plan* should include details of the process communication plan as well as process performance experience, skills, and requirements.
g. Resources and information necessary for performing the process are identified, made available, allocated, and used.	*GP 2.1.7 Identify and make available resources to perform the process according to plan.* Resources and information necessary for performing the key activities of the process are identified, made available, allocated, and used.	*GWP 2.0 Process plan* should provide details of the process training plan and process resourcing plan.
h. Interfaces between the involved parties are managed to ensure effective communication and clear assignment of responsibility.	*GP 2.1.8 Manage the interfaces between involved parties.* The individuals and groups involved with the process are identified, responsibilities are defined, and effective communication mechanisms are in place.	*GWP 1.0 Process documentation* should provide details of the individuals and groups involved (suppliers, customers and RACI). *GWP 2.0 Process plan* should provide details of the process communication plan.

PA 2.2 Work Product Management

This process attribute measures how well the work products produced by the process are managed. Full achievement of this attribute results in the process achieving the following attributes (see Table 4.3).

Table 4.3 Process Attribute 2.2 Work Product Management

RESULT OF FULL ACHIEVEMENT OF THE ATTRIBUTE	GENERIC PRACTICES (GPs)	GENERIC WORK PRODUCTS (GWPs)
a. Requirements for the work products of the process are defined.	*GP 2.2.1 Define the requirements for the work products.* This should include content structure and quality criteria.	*GWP 3.0 Quality plan* should provide details of quality criteria and work product content and structure.
b. Requirements for documentation and control of the work products are defined.	*GP 2.2.2 Define the requirements for documentation and control of the work products.* This should include identification of dependencies, approvals and traceability of requirements.	*GWP 1.0 Process documentation* should provide details of controls; for example, a control matrix. *GWP 3.0 Quality plan* should provide details of work product, quality criteria, documentation requirements and change control.
c. Work products are appropriately identified, documented and controlled.	*GP 2.2.3 Identify, document and control the work products.* Work products are subject to change control, versioning and configuration management as appropriate.	*GWP 3.0 Quality plan* should provide details of work product, quality criteria, documentation requirements and change control.
d. Work products are reviewed in accordance with planned arrangements and adjusted as necessary to meet requirements.	*GP 2.2.4 Review and adjust work products to meet the defined requirements.* Work products are subject to review against requirements in accordance with planned arrangements, and any issues that arise are resolved.	*GWP 4. Quality records* should provide an audit trail of reviews undertaken.

Level 3: Established Process

The process referenced in Level 2: *Managed Process* is now implemented using a defined process that can achieve its process outcomes. To achieve Level 3, the organization must meet the following process attributes.

PA 3.1 Process Definition

This process attribute measures how well standard processes are in place for supporting deployment. Full achievement of this attribute results in the process achieving the attributes shown in section 6 of the book ISO IEC 33063 which ISO Intellectual Property rights restrict us from providing.

PA 3.2 Process Deployment

This process attribute measures how well you deployed and defined the standard process so it can achieve its process outcomes. Full achievement of this attribute results in the process achieving the attributes shown in section 6 of the book ISO IEC 33063 which ISO Intellectual Property rights restrict us from providing.

Level 4: Predictable Process

The process referenced in Level 3: *Established Process* now operates effectively within defined limits to achieves its process outcomes. In Level 4, quantitative management needs are identified, measurement data are collected and analyzed to identify and take action against designated causes of variation. To achieve this level, the organization must meet the following process attributes.

PA 4.1 Quantitative Analysis

This process attribute measures how well information needs are defined, relationships between process elements are identified and data are collected. Full achievement of this attribute results in the process achieving the attributes shown in section 6 of the book ISO IEC 33063 which ISO Intellectual Property rights restrict us from providing.

PA 4.2 Quantitative Process Control

This process attribute measures how well objective data are used in managing predictable process performance. Full achievement of this attribute results in the process achieving the attributes shown in section 6 of the book ISO IEC 33063 which ISO Intellectual Property rights restrict us from providing.

Level 5: Innovating Process

The process referenced in Level 4: *Predictable Process* is now continually improved to respond to change aligned with organizational goals. To achieve Level 5, the organization must meet the following process attributes.

PA 5.1 Process Innovation

This process attribute measures how well changes to the process are identified from investigating how to innovate the definition and deployment of the process. Full achievement of this attribute results in the process achieving the attributes shown in section 6 of the book ISO IEC 33063 which ISO Intellectual Property rights restrict us from providing.

PA 5.2 Process Innovation Implementation

This process attribute measures how well changes in definition, management, and performance helps achieve the innovation objectives. Full achievement of this attribute results in the process achieving the attributes shown in section 6 of the book ISO IEC 33063 which ISO Intellectual Property rights restrict us from providing.

Rating Scale

From ISO/IEC 33020, we get the process attribute rating scales. As stated above, we will measure process attributes using a scale: the scale shown in Table 4.4.

Table 4.4 Process Attribute Ratings

RATING	MEANING	LEVEL OF ACHIEVEMENT
N	Not achieved	0% to ≤15%
P	Partially achieved	>15% to ≤50%
L	Largely achieved	>50% to ≤85%
F	Fully achieved	>85% to ≤100%

Table 4.5 Partially and Largely (P and L) Process Attributes

RATING	MEANING	LEVEL OF ACHIEVEMENT
P–	Partially achieved –	>15% to ≤32.5%
P+	Partially achieved +	>32.5% to ≤50%
L–	Largely achieved –	>50% to ≤67.5%
L+	Largely achieved +	>67.5% to ≤85%

These ratings were available in ISO 15504, but as many who performed assessments will attest, they were a little restrictive. To assist assessors, ISO/IEC 33020 provides a refinement of Partially and Largely (P and L) ratings as shown in Table 4.5.

However, there is still an issue the authors feel, with adequately defining one level from the next. For example, how do you decide between 50 or 51 percent? Our opinion is, when the data is not adequate enough to definitely determine at this detailed level, the assessor should default to the lower level for the assessment.

NOTE

You must begin an assessment by starting with the achievement of Level 1, which is performance of the process. To achieve Level 1, you must rate Level 1 as either Largely or Fully achieved. To achieve subsequent levels, you must fully achieve the preceding level. You may assess higher levels but may only achieve a higher level when all lower levels are rated fully achieved.

So, now you have all the components of the process assessment model. All, you need now is the assessment methodology. In Part II, you will learn how to use these ratings.

PART II
PROCESS ASSESSMENT METHOD

In Part I, you were introduced to the process dimension. In Part II, we introduce you to the capability dimension. Specifically, we present the process capability indicators related to the nine process attributes associated with the five capability levels or Levels 1 through 5.

5

Executing the Assessment—Assessor Guide

In this part, you will find the steps and approach to performing an assessment—ensuring the correct roles and responsibilities, ascertaining independence and competency, and following the specified steps in performing the assessment. When using this book, you are helping ensure assessment results are objective, consistent, repeatable, and representative of the assessed processes.

The book allows you to perform an assessment with one of several goals in mind. These can be used by or on behalf of an organization to:

a. Facilitate self-assessment.
b. Provide a basis for improving process performance and mitigating process-related risk.
c. Produce a rating of the achievement of the relevant process quality characteristic.
d. Provide an objective benchmark between organizations.

Organizations of all sizes can use this method. The purpose of process assessment in this book is to understand and assess the processes involved in project management (PM) within your organization. The requirements within are written to help you ensure that the assessment

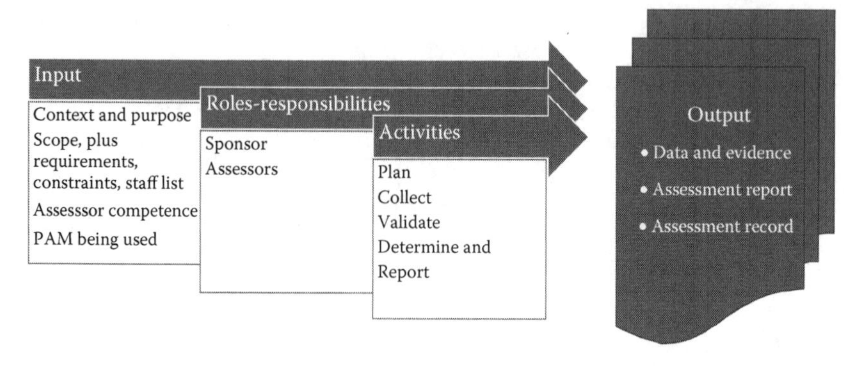

Figure 5.1 Assessment process. You can also find a detailed list of all the steps needed to perform an assessment in Appendix J: Key Steps in an Assessment.

output is consistent and provides evidence to substantiate the ratings decided upon during execution of the assessment.

Figure 5.1 shows the key elements of the assessment process.

Using this book ensures that the assessment process meets the purpose for the assessment; this book is structured to help ensure that the purpose for performing the assessment is satisfied, both in terms of the rigor and independence of the assessment and its suitability.

In the following sections, you will find descriptions of activities and tasks designed to meet the various ISO standards that must be followed to ensure a compliant assessment. A later section will let you know how to conduct a less rigorous self-assessment that is likely easier to accomplish as it does not require validated evidence and is essentially done at a higher level of detail than a formal assessment.

KEY SUCCESS FACTORS

Ensuring sponsor support for the assessment
Adequately defining the scope of the assessment
Ensuring assessors understand the process of assessment
Ensuring appropriate project management expertise

Requirements of the Assessment

In terms of the strategy for collecting and analysing data, each assessment must identify the assessment activities, roles, responsibilities and competences, classes of assessment used, and the rating and aggregation methods. Each assessment must also define the criteria for ensuring coverage for both the defined organizational scope and the defined process scope for the assessment. Following the steps outlined in this book will help ensure the above is accomplished.

Assessment Steps

The diagram that follows shows a simple Swim-Lane for performing an assessment.

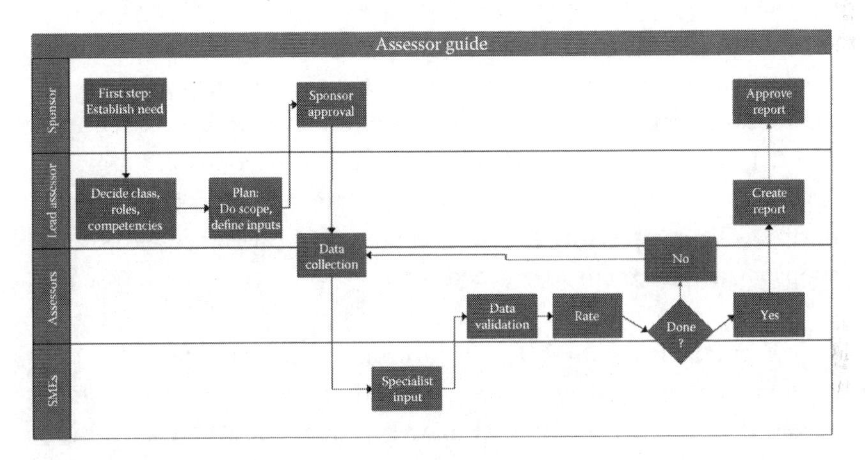

There are several requirements for ensuring your assessment is performed according to the specifications of the ISO/IEC 33000 series. These are all set out in the following pages. It is intended that you are able to perform an assessment by following these steps and referring to the associated Work Papers and other sections of this book for additional detail. Steps three through five are repeated until all data needed for the assessment are collected (see Figure 5.2).

Each step is performed in order with steps 3 and 4 being possibly done multiple times in order to complete all the Process Instances necessary for each Level. You may perform Ratings as you validate the data rather than waiting until the data is collected for the Level you are performing; thus, you will see in Figure 5.2 that those levels have been combined.

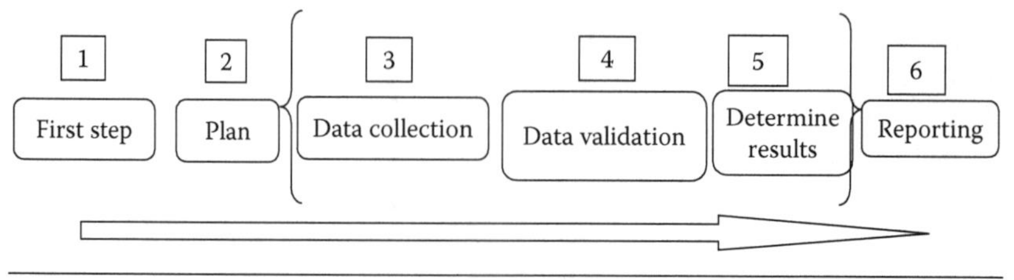

Figure 5.2 Assessment steps.

The First Step

The first step is determining a need for an assessment of your project management processes. Typically, this comes from some sort of realization that your projects are perhaps not performing at the level you wish them to perform. Alternatively, you may wish to assess the present capability of your PM system and processes. For either, this decision forms part of the start-up reason and needs to be ascertained to ensure you obtain the necessary results. Remember, the basic purpose of process assessment is to understand and assess the processes implemented within your PM methodology. You also need to determine the class of assessment to ensure you follow all the necessary requirements during your assessment. The assessment begins with the assessment sponsor's commitment to proceed.

Classes of Assessment When performing an assessment, the purpose for the assessment can vary depending on the type of business involved and the context within the organization being assessed. To help ensure a high degree of success, three distinct classes of assessment are defined. Those three classes provide different levels of confidence. You need to determine which class best fits your needs. For example, if you need a high degree of confidence and the ability to ensure the results provide a good representation of process performance across the organization, you might need to use a Class 1 assessment; however, it is unlikely your organization would begin assessing at this level. It is far more common to perform Class 2 or 3 assessments due the level of work involved and to see where you organization stands.

NOTE

Use of a Class 3 assessment to provide a rating of organizational process maturity is not recommended as it is more suitable for monitoring the ongoing progress of an improvement program, or identifying key issues for a later Class 1 or Class 2 assessment.

Each class results in a different level of confidence. You need to ensure the one you choose provides the degree of confidence needed by your organization. This confidence level may differ depending upon whether you are assessing the overall PM methodology in use or assessing whether one particular project is being performed at a suitable level of capability. Table 5.1 describes each of the classes and their specific requirements.

Table 5.1 Classes of Assessment

CLASS	PURPOSE	REQUIREMENTS
Class 1	1. Provide a level of confidence in the results of the assessment such that the results are suited for comparisons across different organizations 2. Enable assessment conclusions to be drawn as to the relative strengths and weaknesses of the organizations compared 3. Provide a basis for process improvement, external benchmarking and process quality determination	1. Category of independence of all members of the assessment team shall be recorded 2. At least one assessor is the Lead Assessor 3. A minimum of four process instances (where possible) shall be identified for each process within the scope of the assessment
Class 2	1. Provide a level of confidence in the assessment results that may indicate the overall level of performance of the key processes in the organization unit, which are suitable for comparisons of the results of an assessment across an organizational or product line scope 2. Enable assessment conclusions to be drawn about the opportunities for improvement and levels of process-related risk 3. Provide a basis for an initial assessment at the commencement of an improvement program	1. Category of independence of all members of the assessment team shall be recorded 2. At least one assessor is the Lead Assessor 3. A minimum of two process instances (where possible) shall be identified for each process within the scope of the assessment

(*Continued*)

Table 5.1 (*Continued*) Classes of Assessment

CLASS	PURPOSE	REQUIREMENTS
Class 3	1. Generate results that may indicate critical opportunities for improvement and key areas of process-related risk 2. Be suitable for monitoring the ongoing progress of an improvement program, or to identify key issues for a later Class 1 or Class 2 assessment	1. There are no specific requirements for a Class 3 assessment

Source: Permission to use extracts from ISO/IEC 33002:2015 was provided by the Standards Council of Canada (SCC). No further reproduction is permitted without prior written approval from SCC.

Additional, specific requirements will assist you in deciding the particular class you need. In the paragraphs below, you will read about the key elements of those added requirements to help you make a preliminary decision about which class you will select for your assessment.

Class 1 Additional Requirements For Class 1 assessments, specific requirements are defined relating to the planning, collection and validation of data, attribute rating, and recording the level of the assessment team's independence. During the planning phase, you need to record the level of independence of the assessment team. You will need to assign at least one assessor, the Lead Assessor, in this case. Additionally, you must plan to identify at least four (4) process instances[1] for each process you put in scope. Use them all if there are fewer than four.

When collecting and validating the data, for each process and process attribute outcome in scope, you need to collect objective evidence from the evaluation of work products and testimony of process performers.

When determining and reporting the results, you need to, based on validated data, characterize the extent to which each process and process attribute outcome is achieved for each process instance. Every process attribute within scope will be rated, based on validated data. If a process attribute cannot be rated, then it will be documented

[1] For example, in a business process management solution, process instances can be started using events such as submitting a form, sending an email, or creating or uploading files in a shared drive, such as Microsoft SharePoint. Each process instance can be independently monitored and managed.

as a gap in performance. The assessment team will judge whether a gap represents an overall weakness, ensuring that gaps and weakness statements are documented and retained. Once all ratings are assigned, the team may determine the maturity level, if desired, based on an acceptable maturity model. Finally, the report will be approved by the lead assessor, confirmed by all team members, and authorized by the body performing the assessment.

Class 2 Additional Requirements For Class 2 assessments, specific requirements are defined relating to the planning, collection and validation of data, attribute rating, and recording the level of the assessment team's independence. During the planning phase, you need to record the level of independence of the assessment team. You will need to assign at least one assessor, the Lead Assessor in this case. Additionally, you must plan to identify at least two (2) process instances for each process you put in scope. Use the available instances if there are fewer than two.

NOTE

In Class 2 assessments, it is recommended that the lead assessor be independent of the organizational unit being assessed.

When collecting and validating the data, for each process and process attribute outcome in scope, you need to collect objective evidence from the evaluation of work products and testimony of process performers.

When determining and reporting the results, you need to, based on validated data, characterize the extent to which each process and process attribute outcome is achieved for each process instance. Every process attribute within scope will be rated based on validated data. If a process attribute rating for the highest attribute rating of a given process instance cannot be accomplished, then it will be documented as a gap in performance. Following completion of ratings, the team will determine the set of process profiles and quality levels. Once all ratings are done the team may determine the maturity level if wanted, based on an acceptable maturity model. Finally, the report will be approved by the lead assessor, confirmed by all team members, and authorized by the body performing the assessment.

Class 3 Specific Requirements No additional requirements are specified.

Assessor Independence There are several considerations regarding the need for an independent assessor. The bodies performing the assessment and the team members involved can be categorized into three different types. You will select the particular category needed for the specific assessment being undertaken. You will find the details for each category below (also summarized in Table 5.2). Which category you decide to use will be based upon the factors involved within each class and your particular business needs.

Category A Organizations in Category A provide fully independent, third party services.

Category B Category B includes an internal team of assessors within the organization that provide assessments with a lead assessor from an independent organization. This type of assessment can be used in a verification-based approach where data is collected by internal team members.

Category C This may be an internal group with a separate reporting line, that is, independent, from the area being assessed. It typically is used by a large organization with a separate functional group responsible for performing assessments.

Category D In this category, the assessor may be an internal consultant that is assisting an organization in implementing process improvement which then assesses its capabilities. This category might be more applicable to small organizations with little or no pressure from customers for an independent assessment.

Roles and Responsibilities You also need to ensure the roles and responsibilities are spelled out and clear. Depending upon the class of assessment you choose, you may need one or more assessors. You also need to identify those in the organization who are needed to assist in the assessment. Use the following and Table 5.2 to help determine the team members needed for your assessment.

Table 5.2 Categories of Independence of Bodies and Personnel Performing Assessment

	CATEGORY A	CATEGORY B	CATEGORY C	CATEGORY D
Body performing the assessment	The body performing the assessment shall be independent of the organization being assessed.		The body performing the assessment shall be part of the organization being assessed.	
Lead assessor	The lead assessor shall be independent of the organization being assessed.		There shall be adequate separation of responsibilities of the lead assessor from personnel in the organizational unit being assessed.	The lead assessor can be part of the organizational unit being assessed.
Assessors (other than lead assessor)	The assessors shall be independent of the organization being assessed.	There shall be a separation of the responsibilities of the assessors from personnel in the organizational unit being assessed.	There shall be a separation of the responsibilities of the assessors from personnel in the organizational unit being assessed.	The assessors can be part of the organizational unit being assessed.

Source: Permission to use extracts from ISO/IEC 33002:2015 was provided by the Standards Council of Canada. No further reproduction is permitted without prior written approval from SCC.

The Team An assessment needs an assessor or a Lead assessor depending upon the assessment class and the category of independence in use (refer to Classes of Assessment, Assessor Independence, Table 5.1, and Table 5.2 for details). Remember, most assessments will likely begin with a Class 3 or Class 2 assessment, because the work involved in completing a Class 1 assessment is more onerous and may only be worth doing when lower level assessments show the need or you intend to do comparisons across different departments or organizations. The number of assessors involved will be primarily dependent on the size of the organization being reviewed and the number of processes involved.

Sponsor The sponsor of the assessment needs to verify that the individual who is taking responsibility and is designated as the lead assessor has the required competences. The sponsor also needs to finalize the assessment scope and approve the assessment plan. They need to make sure resources are made available as needed and that the assessment team has access to any relevant resources. You want this person to be one with significant authority within the organization to help ensure that organizational politics and any scheduling or other issues are easily resolved.

Lead Assessor The lead assessor confirms the sponsor's commitment, understands and documents the sponsor's assessment objectives and verifies that the planned approach follows all requirements. They verify that the planned assessment scope is the same as the desired one and ensures that participants in the assessment are properly briefed on the purpose, scope and assessment approach, ensuring that all members of the assessment team have the knowledge and skills appropriate to their role on the project. They also ensure everyone on the team has access to appropriate documented guidance on how to perform the assessment activities, and everyone on the team has the competences to use any tools chosen to support the assessment. It is the lead assessors job to confirm receipt of the assessment result deliverables by the sponsor and, upon completion of the assessment, to verify and document the extent of conformance to these requirements.

Assessors Assessor(s) carry out assigned activities such as detailed planning, data collection, data validation, rating the process attributes and reporting. They need to be competent with appropriate education, training and experience, including domain experience, to perform the required class of assessment and make professional judgements. They should have experience with project management as those are the processes involved. Ideally, they may even hold a certification, such as a Project Management Professional (PMP®), although this is not a firm requirement. Regardless, they need to be fully qualified in project management to adequately assess the processes involved.

Coordinator This position comes from within the organization and needs to be someone who can ensure that the right people are involved and understand their roles, meetings are scheduled and attended, enquiries are responded to appropriately and any needs of the assessment team are met. When your assessment is large enough or a Class 1 assessment, you may find a coordinator is necessary to coordinate meetings and receipt of artefacts.

Competency The team members must be competent on the basis of appropriate education, training and experience, including domain experience, to perform the required class of assessment and make professional judgements. The sponsor ensures that the Lead Assessor is capable while that person then ensures all assessors are capable and able.

This means team members are people who understand and are deeply familiar with the capability assessment process, project management processes, evidence collection procedures, data collection and marking and interviewing skills.

KEY STEPS

Determine the business need
Determine the class of assessment
Get sponsor approval
Determine the category of independence needed
Establish roles and responsibilities
Assess competency requirements

Plan

With the sponsor committed to the assessment, a plan needs to be prepared and distributed that includes the assessment inputs, the class of assessment and category of independence, communications to the staff involved, the activities to be performed and the resources to be used along with their schedule.[2]

As well, the participants roles and responsibilities that were identified in the earlier step need to be defined, and the assessment output properly described. You might, in this phase, decide to initiate a pre-assessment questionnaire to those involved. In it you might ask general questions about the processes being covered, the amount of time available for this assessment, how the current project methodology works, any issues that are known and the level of understanding the assessed staff have of capability assessments. The plan needs documented approval from the sponsor. In addition, you must outline and agree upon the strategy and techniques for the selection, identification, collection and analysis of objective evidence and data necessary to satisfy any requirements for coverage of the organizational scope or the process scope of the assessment as defined for the class of the assessment.

So how are you going to decide which project management processes you will assess? Deciding on the scope is a crucial step as too many processes and your assessment will likely flounder under the volume of work.

Scoping the Assessment Earlier in the book, it was noted that while you will do your assessment based on the processes that support the organizational goals and objectives, any initial assessment should focus on the core planning processes. "If you fail to plan, you plan to fail" is entirely appropriate in project management. In reality, you should spend 80% of your time in planning and 20% in execution. Therefore, it is recommended that your initial assessment involve the Planning domain and associated processes.[3]

---

[2] Remember the assessment itself is a project. A class 1 in your approach can be a major project if it spans the entire organization such as one for a "Fortune" 100 company.

[3] That is assuming you have reviewed ES01: *Define the project management framework* and ES02: *Set policies, processes and methodologies.*

In this key Domain, there are sixteen (16) specific processes. You may want to prioritize which of these provide the greatest value as doing all 16 processes may be too large an assessment. You could prioritize using your knowledge and expertise and the Pareto principle of 80–20 where 20% of the processes provide 80% of the value. Do these 20% first (Figure 5.3).

However, that is just one way to scope your assessment. You might also do a review of your previous projects and determine which domain provided the most difficulty in those projects and choose that for the assessment. A third method would be to decide which of the seven domains is most important to your business objectives and begin there. For example, does the organization have difficulty deciding whom to assign to each project? If so, then the Initiate domain may a good starting point for assessment. Likewise, should the projects tend to run out of steam and begin to drift, then the Implement domain may be best. Last, but not least, perhaps your internal and external auditors keep hammering you on a project management process or lack of specific work products, which is now a pain point. Then, this pain point would be an excellent place to start. Regardless, with seven domains and 45 processes you need to focus your assessment in some manner. Trying to assess the entire project management methodology at the same time would likely be asking for failure because of the work involved. As a rule of thumb, pick no more than three domains when starting. Your organization is not likely to absorb change in more than three processes at any given time.

Assessing Each Attribute You will begin your assessment by first assessing the Level 1 attribute, Process Performance. Here, you will use the individual Process descriptions shown earlier for the particular project management processes you are including in your project scope. You will determine whether there is a process using the Process Purpose statement. Then, you need to determine whether the Process Outcomes are achieved. You do this by reviewing the Base Practices involved and assessing their level of compliance. This involves ensuring that the appropriate Work Products are used as input and that the needed Work Product Outputs are created. This is the only time you refer to the processes as defined in the above PAM. You can see this in the following diagram showing one of the Processes involved (Table 5.3).

Governance

Establish	ES01 Define the project management framework ES02 Set policies, processes and methodologies ES03 Set limits of authority for decision-making	Monitor	MO01 Ensure project benefits MO02 Ensure risk optimization MO03 Ensure resource optimization

Management

Initiate	IN01 Develop project charter IN02 Identify stakeholders	IN03 Establish project team
Plan	PL01 Develop project plans PL02 Define scope PL03 Create work breakdown structure PL04 Define activities PL05 Estimate resources PL06 Define project organization PL07 Sequence activities PL08 Estimate activity durations	PL09 Develop schedule PL10 Estimate costs PL11 Develop budget PL12 Identify risks PL13 Assess risks PL14 Plan quality PL15 Plan procurements PL16 Plan communications
Implement	IM01 Direct project work IM02 Manage stakeholders IM03 Develop project team IM04 Treat risks	IM05 Perform quality assurance IM06 Select suppliers IM07 Distribute information
Control	CO01 Control project work CO02 Control changes CO03 Control scope CO04 Control resources CO05 Manage project team CO06 Control schedule	CO07 Control costs CO08 Control risks CO09 Perform quality control CO10 Administer procurements CO11 Manage communications
Close	CL01 Close project phase or project	CL02 Collect lessons learned

Figure 5.3 Process reference model.

Table 5.3 Assessing the Level 1 Attribute

Area	Governance	
Domain	Establish	
Process ID	ES01	
Process Name	Define and manage the project management framework	
Process Description	Clarify and maintain the governance framework for project management. Implement and maintain tools and techniques and authorities to manage projects in line with enterprise objectives.	
Process Purpose Statement	The purpose of *Define the project management framework* is to provide a consistent project management approach to enable enterprise governance requirements to be met, covering project management processes, organizational structures, roles and responsibilities, reliable and repeatable activities, and skills and competencies.	
OUTCOMES		
NUMBER	DESCRIPTION	
ES01-01	A project management framework is defined.	
ES01-02	A project management framework is followed.	
BASE PRACTICES (BPs)		
NUMBER	DESCRIPTION	SUPPORTS
ES01-BP 1	Define a project management framework for IT investments.	ES01-01
S01-BP 2	Establish and maintain an IT project management framework	ES01-01:02
ES01-BP 3	Establish Project Management Office	ES01-01

Is this Process in use?

Are you achieving these outcomes?

(*Continued*)

Table 5.3 (*Continued*) Assessing the Level 1 Attribute

WORK PRODUCTS			
INPUTS			
NUMBER	DESCRIPTION		SUPPORTS
From corporate governance	Decision-making model		ES01-01:02
ES03-WP1	Authority limits		ES01-01:02
From corporate governance	Enterprise governance guiding principles		ES01-01:02
From corporate governance	Process architecture model		ES01-01
OUTPUTS			
NUMBER	DESCRIPTION	INPUT TO	SUPPORTS
ES01-WP1	Project management guidelines	All processes	ES01-01
ES01-WP2	Project management office Charter	Project Manager	ES01-01
ES01-WP3	Definition of organizational structure and functions		
ES01-WP4	Definition of project-related roles and responsibilities		
ES01-WP5	Process capability assessments		ES01-02
ES01-WP6	Performance goals and metrics of process improvement tracking		ES01-02

Are all these inputs and outcomes being produced?

For assessments beyond Level 1, you use Generic Attributes, rather than the specific ones in Level one. These Generic Attributes apply regardless of the particular process you are assessing. A list of all these are found in the section titled—Process Capability Indicators.

For instance, if you intend to move to assessing Level 2 once your assessment of Level 1 is complete, then there are two indicators involved. Both have to be satisfied in order for your assessment to be complete at that level. In order to achieve that level of course, Level 1 needs to be Largely or Fully complete.

At this level, the generic Attributes consist of PA 2.1 Performance Management and PA 2.2 Work Product Management. You can see in the diagram below that there are several Generic Practices and Work products for you to review.

NOTE

There is no particular need to be at one Level or another. This is determined by the needs of your organization and the level of effort and cost involved. You should, however, ensure you have reached Level 1. You may wish to attain higher levels for more important processes.

The first column is indicating what the overall result of achievement involves. To indicate achievement of the attribute, all the Generic Practices in the column two along with the associated Generic Work Products in column three need to achieved for the Process you are assessing. As you can see, rather than the specifics of the earlier Level 1 that apply directly to the process you are assessing, these same generic items apply to any of the project management processes you choose to assess.

Example 5.1: PA 2.1 Performance Management

This process attribute is a measure of the extent to which the performance of the process is managed. Full achievement of this attribute results in the process achieving the following attributes (Table 5.4).

Table 5.4 PA 2.1 Performance Management

RESULT OF FULL ACHIEVEMENT OF THE ATTRIBUTE	GENERIC PRACTICES (GPS)	GENERIC WORK PRODUCTS (GWPS)
a. Objectives for the performance of the process are identified.	GP 2.1.1 Identify the objectives for the performance of the process. The performance objectives, scoped together with assumptions and constraints, are defined and communicated.	GWP 1.0 Process documentation. Should outline the process scope. GWP 2.0 Process plan. Should provide details of the process performance objectives.
b. Performance of the process is planned.	GP 2.1.2 Plan the performance of the process to fulfill the identified objectives. Basic measures of process performance linked to business objectives are established. They include key milestones, required activities, estimates and schedules.	GWP 2.0 Process Plan. Should provide details of the process performance objectives. GWP 9.0 Process performance records. Should provide details of the outcomes.
c. Performance of the process is monitored.	GP 2.1.3 Monitor the performance of the process to fulfill the identified objectives. Basic measures of process performance linked to business objectives are established and monitored. They include key milestones, required activities, estimates and schedules.	GWP 2.0 Process Plan. Should provide details of the process performance objectives. GWP 9.0 Process performance records. Should provide details of the outcomes.
d. Performance of the process is adjusted to meet plans.	GP 2.1.4 Adjust the performance of the process. Action is taken when planned performance is not achieved. Actions include identification of process performance issues and adjustment of plans and schedules as appropriate.	GWP 4.0 Quality record. Should provide details of action taken when performance is not achieved.

What you expect to occur

The things you expect that support the Result

The types of items that should be in place to achieve the GP

(*Continued*)

Table 5.4 (*Continued*) PA 2.1 Performance Management

What you expect to occur

The things you expect that support the Result

The types of items that should be in place to achieve the GP

RESULT OF FULL ACHIEVEMENT OF THE ATTRIBUTE	GENERIC PRACTICES (GPS)	GENERIC WORK PRODUCTS (GWPS)
e. Responsibilities and authorities for performing the process are defined, assigned and communicated.	GP 2.1.5 Define responsibilities and authorities for performing the process. The key responsibilities and authorities for performing the key activities of the process are defined, assigned and communicated. The need for process performance experience, knowledge and skills is defined.	GWP 1.0 Process documentation. Should provide details of the process owner and who is responsible, accountable, consulted and informed (RACI). GWP 2.0 Process plan. Should include details of the process communication plan as well as process performance experience, skills and requirements.
f. Personnel performing the process are prepared for executing their responsibilities.	GP 2.1.6 Assign personnel and train them to perform the process according to plan. Personnel performing processes accept responsibility and have the necessary skills and experience to carry out their responsibilities.	GWP 2.0 Process plan. Should include details of the process communication plan as well as process performance experience, skills and requirements.
g. Resources and information necessary for performing the process are identified made available, allocated and used.	GP 2.1.7 Identify and make available resources to perform the process according to plan. Resources and information necessary for performing the key activities of the process are identified, made available, allocated and used.	GWP 2.0 Process plan. Should provide details of the process training plan and process resourcing plan.
h. Interfaces between the involved parties are managed to ensure effective communication and clear assignment of responsibility.	GP 2.1.8 Manage the interfaces between involve parties. The individuals and groups involved with the process are identified, responsibilities are defined and effective communication mechanisms ae in place.	GWP 1.0 Process documentation. Should provide details of the individuals and groups involved (suppliers, customers and RACI). GWP 2.0 Process plan. Should provide details of the process communication plan.

Example 5.2: PA 2.2 Work Product Management

This process attribute is a measure of the extent to which the work products produced by the process are appropriately managed. Full achievement of this attribute results in the process achieving the attributes shown in Table 5.5.

The example above applies across all processes being assessed at Level 2, and as needed the remaining Generic Processes for the higher Levels 3 through 5 would apply for all processes that you wish to assess at those higher levels. Remember, there is no particular need to be at any given level other than Level 1. After all, why have processes when they do not actually perform as desired?

Table 5.5 PA 2.2 Work Product Management

RESULT OF FULL ACHIEVEMENT OF THE ATTRIBUTE	GENERIC PRACTICES (GPS)	GENERIC WORK PRODUCTS (GWPS)
a. Requirements for the work products of the process are defined.	GP 2.2.1 Define the requirements for the work products. This should include content structure and quality criteria.	GWP 3.0 Quality plan. Should provide details of quality criteria and work product content and structure.
b. Requirements for documentation and control of the work products are defined.	GP 2.2.2 Define the requirements for documentation and control of the work products. This should include identification of dependencies, approvals, and traceability of requirements.	GWP 1.0 Process documentation. Should provide details of controls; for example, control matrix. GWP 3.0 Quality plan. Should provide details of work product, quality criteria, documentation requirements, and change control.
c. Work products are appropriately identified, documented, and controlled.	GP 2.2.3 Identify, document and control the work products. Work products are subject to change control, versioning, and configuration management as appropriate.	GWP 3.0 Quality plan. Should provide details of work product, quality criteria, documentation requirements, and change control.
d. Work products are reviewed in accordance with planned arrangements and adjusted as necessary to meet requirements.	GP 2.2.4 Review and adjust work products to meet the defined requirements. Work products are subject to review against requirements in accordance with planned arrangements and any issues arising are resolved.	GWP 4. Quality records. Should provide and audit trail of reviews undertaken.

To go a higher level is a business decision based on your need and desire to expend the additional effort needed to maintain a process at each successive higher level. You would do so when the process directly supports achievement of your organizations' objectives.

Assessment Inputs In this phase, you will also define and manage the assessment inputs needed for a successful project, ensure that any changes made are approved by the sponsor or their delegate and the changes are adequately documented.

At minimum, the assessment inputs have to specify:

1. Who the sponsor is and their role in the area
2. The business context reason for the assessment
3. The assessment scope, including:
   a. Processes to be investigated within each organizational unit
   b. The size, criticality, complexity of the project
4. Requirements including:
   a. The assessment class and levels of independence and the ratings used
5. Constraints considering, at minimum:
   a. Availability of key resources
   b. Maximum duration of the assessment
   c. Specific processes to be excluded from the assessment
   d. Ownership of the assessment outputs and any restrictions on their use
   e. Handling of confidential information and any need for non-disclosure.
6. Identity and roles of everyone involved
7. Criteria for competence of the lead assessor. Finally, remember to ensure that your assessment record is clear and concise. It should also include:
   a. Date of the assessment
   b. The assessment inputs gathered
   c. The objective evidence gathered related to your findings
   d. Identification of the documented assessment process
   e. The resulting process profiles
   f. The capability levels achieved

KEY STEPS

Document the class of assessment and category of independence
Determine the assessment scope
Determine communications to the staff involved
Set out the activities to be performed
Assign the resources to be used
Determine resource schedules
Document assessment inputs
Describe assessment outputs
Decide whether to initiate a pre-assessment questionnaire
Describe the strategy and techniques for the selecting, identifying, collecting and analyzing objective evidence and data
Document sponsor approval of the plan

Data Collection

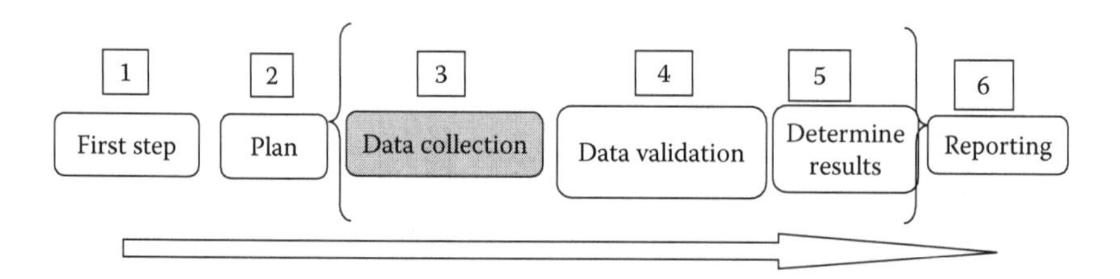

In this phase assessors collect relevant information using interviews and available data such as policies, reports, procedures, and any other useful data required to determine adequately the degree of compliance. It is always best to include two assessors when interviewing subjects since information may be missed while note taking. The second assessor is could listen and add additional questions as needed. Open-ended questions help the assessor pull information from the interviewee. For example, "What issues have you had with this process over the last month" might be a better question than "Is the process working okay?"

When collecting data, the data need to be sufficient to provide coverage of the organization and process scope for the assessment, as specified when you selected the particular assessment class. Within the class, you need to collect the data using direct or indirect evidence that satisfies the particular class of assessment. Direct evidence would be written project plans, the project charter, and the specific PM framework in use. Indirect evidence would be interview notes and discussions with project participants.

Evidence Requirements You need to collect the required evidence in a systematic manner with the points provided below used at a minimum:

a. Assess each process in the assessment scope on the basis of objective evidence.
b. Identify and gather the objective evidence to enable reasonable level of assurance.
c. Ensure the evidence meets the needs of the assessment purpose, scope and class.
d. Make sure the information is relevant to understanding of the assessment output.

It would be a good idea to determine a schema for identifying relevant data corresponding to the particular attribute it belongs to satisfy the traceability requirements of ISO/IEC 33000 series. ISO has no specified instruction on how you might accomplish this, only that you must relate each piece of validated evidence to the specific attribute that it relates so as to allow other assessors ready access should the sponsor or someone else question an assessment.

Feel free to choose whichever method you prefer; however, the following can serve as a guideline.

Each Domain has been assigned a number and the corresponding processes with a Process ID. Within each process in that domain, each Outcome, Base Practice and Work Product are also numbered.

Use these numbers to reference any materials collected, whether interview notes or documents with these numbers and a further number identifying the actual piece of evidence.

So, for example, if you are assessing the Plan domain, and the *Define Scope* process you will start with PL02. Next, if you are documenting the Base Practice 1—*Determine Project Scope* then BP 1 would follow, giving you PL02-BP 1. Your first piece of evidence might then be labelled PL02-BP 1-1, followed by PL02-BP 1-2 and onward. Assessing a different process such as Plan—*Define Activities* would begin with PL04 followed by the particular WP or Base Practice. This enables you to easily ensure traceability by identifying all interview notes and documents for later review of any specific aspect of the overall assessment. See Appendix H Sample Tracking Form for an example of how you might track materials supporting your assessment.

KEY STEPS

Set up key interviews
Determine record traceability plan
Review collected data
Follow evidence requirements

Data Validation

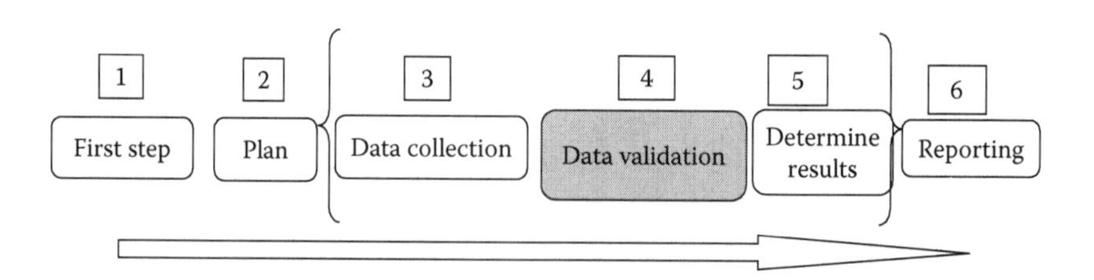

The data from each instance then need to be properly validated. It has to be objective, sufficient and representative enough to cover the assessment purpose. The data should also cover the organization scope and assessment class and finally be consistent as a whole. How does this translate into reality? How do you "validate" something?

The Merriam-Webster dictionary indicates the following:

> Check or prove the validity or accuracy of (something):
> "these estimates have been validated by periodic surveys"
> demonstrate or support the truth or value of:
> "in a healthy family a child's feelings are validated"
> **Synonyms:** prove substantiate corroborate verify support back up bear out lend force to confirm justify vindicate authenticate

You should be striving to make sure the document or interview is worthwhile, accurate, and relevant to the information you need. This means the information you gather will help prove the PM process attributes support the achievement you are anticipating. In PL02-BP 1 *Determine project scope*, you should be looking to find a detailed scope statement for the project.

Make sure this is done for the data you collect. This helps ensure the robustness and accuracy of the assessment and helps you eventually rate the particular attribute. Ensure that you collect sufficient data to readily validate data applicability, ability, objectivity, and whether the data is sufficient for the assessment needs.

KEY STEPS

Determine whether the evidence is objective
Determine whether the evidence is sufficient
Decide whether it is representative enough to cover the assessment purpose and class
Determine whether the evidence is consistent as a whole

Determine Results

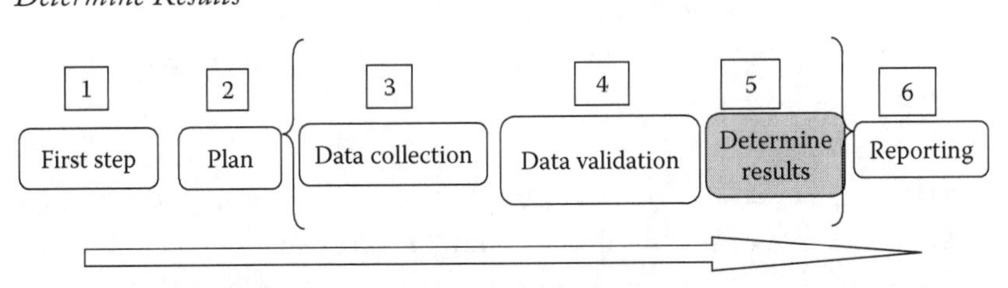

You need to ensure that the defined set of assessment indicators are used to support your judgement as assessors when you analyze the validated data. Recall that the PAM is based on the principle that one could assess the capability of a process by demonstrating the achievement of process attributes on the basis of evidence related to assessment indicators.

There are two types of assessment indicators: process capability indicators, which apply to capability Levels 1 to 5 and process performance indicators, which apply exclusively to capability Level 1. So, if you are doing a Level 1 assessment, the base work products of the particular process you are assessing constitute the set of individual process performance indicators. For higher levels, you use generic process capability indicators.

As well, you need express ratings in the terms used within the measurement framework. The team follows a consistent path that includes rating according to the selected method and decided upon using an approved stakeholder aggregation method. The team needs to maintain traceability between the rating and the evidence used and record the relationship between the assessment indicators chosen for each process attribute and the evidence used as discussed in the earlier section. Then record the profile for the process and, if used, the process quality levels that are within the scope of the assessment. You might, in rare cases, and following earlier assessments, record the maturity level. All these get linked to the assessment purpose and business context such as the target profile or desired outcomes.

To perform your ratings, you can follow the scales indicated in the next section.

Rating Process Capability A process attribute rating is a judgement of the degree of achievement of the process attribute for the assessed process. Attributes are rated to accomplish an agreement on the degree of achievement that a particular process has reached. While the overall rating scale was discussed earlier, below you will learn the requirements used each time you determine the rating of a particular attribute.

Determining the Rating Scale A process attribute is measured using an ordinal scale as defined earlier and shown again below.

The ordinal scale, representing an interval, is more easily used by determining the level of compliance in terms of percentage achievement of a process attribute.

Table 5.6 Rating Scale and Percentage

RATING	MEANING	LEVEL OF ACHIEVEMENT
N	Not achieved	0%–≤15%
P	Partially achieved	>15%–≤50%
L	Largely achieved	>50%–≤85%
F	Fully achieved	>85%–≤100%

Each of the rating scales correspond to a particular percentage scale as shown in Table 5.6

However, in the earlier ISO/IEC 15594 there might be confusion when rating Levels P and L and so should you find it helpful, you may use the refined interval scale for the measures P and L as defined below.

***P+** Partially achieved*: There is some evidence of an approach to, and some achievement of, the defined process attribute in the assessed process. Some aspects of achievement of the process attribute may be unpredictable.

***P–** Partially achieved*: There is some evidence of an approach to, and some achievement of, the defined process attribute in the assessed process. Many aspects of achievement of the process attribute may be unpredictable.

***L+** Largely achieved*: There is evidence of a systematic approach to, and significant achievement of, the defined process attribute in the assessed process. Some weaknesses related to this process attribute may exist in the assessed process.

***L–** Largely achieved*: There is evidence of a systematic approach to, and significant achievement of, the defined process attribute in the assessed process. Many weaknesses related to this process attribute may exist in the assessed process.

The corresponding percentages are shown in Table 5.7.

Table 5.7 Rating Scale and Percentage for Partially and Largely

SCALE	MEANING	PERCENTAGE
P–	Partially achieved minus	>15%–≤32.5%
P+	Partially achieved plus	>32.5%–≤50%
L–	Largely achieved minus	>50%–≤67.5%
L+	Largely achieved plus	>67.5%–≤85%

NOTE

When performing ratings, we believe that should there be a question when the assessor is deciding between one level or the next, such as whether or not an outcome is Partially or Largely achieved because if the validated data is not clear enough to decide unequivocally, then the lower Level must be chosen.

Each Level must be Fully or Largely achieved prior to continuing on to the higher level. There is no Level 2.5 or 3.5. You either achieve the level or you do not. So, when you have Fully achieved all the requirements for Level 1 and are working on Level 2, but have only reached Partial achievement, you remain at Level 1. Once the work is done to achieve Largely or Fully in Level 2 after achieving Fully compliant in Level 1 you can indicate you have achieved Level 2. Table 5.8 outlines the attributes that must be achieved within each Level for you to be able to indicate you have achieved them.

There is no driving need to achieve one level or over another. It is not a race to the top but rather a specified pathway to the desired level of achievement your organization believes necessary for the processes you are assessing. Many organizations will never desire to reach Level 5. The cost of doing this will most likely far exceed any benefits. However, there may be processes where you desire the highest level of achievement. This is a decision each organization must make for each process and will depend upon factors such as the benefit to the organization and desire to compete with other organizations.

Process Rating Methods When you are ready to start rating, you must use a method that is specifically defined relevant to the class of assessment. This can vary according to the class, scope and even the conditions of your particular assessment. The Lead Assessor will need to

Table 5.8 Process Capability Ratings

SCALE	PROCESS ATTRIBUTES	RATING
Level 1	Process Performance	Largely or fully
Level 2	Process Performance	Fully
	Performance Management	Largely or fully
	Work Product Management	Largely or fully
Level 3	Process Performance	Fully
	Performance Management	Fully
	Work Product Management	Fully
	Process Definition	Largely or fully
	Process Deployment	Largely or fully
Level 4	Process Performance	Fully
	Performance Management	Fully
	Work Product Management	Fully
	Process Definition	Fully
	Process Deployment	Fully
	Quantitative Analysis	Largely or fully
	Quantitative Control	Largely or fully
Level 5	Process Performance	Fully
	Performance Management	Fully
	Work Product Management	Fully
	Process Definition	Fully
	Process Deployment	Fully
	Quantitative Analysis	Fully
	Quantitative Control	Fully
	Process Innovation	Largely or fully
	Process Innovation Implementation	Largely or fully

Note: Permission to use extracts from ISO/IEC 33063:2015 was provided by the Standards Council of Canada (SCC). No further reproduction is permitted without prior written approval from SCC.

decide which rating method to use. The selected rating method(s) must then be documented and referenced in the assessment report. There are three (3) rating methods available. Choosing which one to use is the Lead Assessor's task.

Rating Method R1 Within the scope of the assessment, you will describe all process outcomes for as many process instances that are involved based on validated data. Additionally, for all the process attribute (PA) outcomes involved in the assessment, you will also describe each PA based on validated data. Then, you will aggregate the process outcome descriptions for all the assessed process instances to provide a process performance attribute achievement rating. Finally, you will aggregate all the process attribute outcome descriptions providing an overall process attribute achievement rating.

Rating Method R2 With method 2, you will describe each process attribute used in the scope of the assessment for as many process instances that are involved based on validated data. Additionally, you will aggregate all the process attribute descriptions to provide a process attribute achievement rating.

Rating Method 3 No aggregation will take place; so how will these aggregations occur? They can be done across one or two dimensions. Within a process, aggregate the associated process outcomes thus performing a vertical, one-dimension aggregation. If there are multiple process instances, then the outcomes for the associated process or attributes across those instances can be aggregated horizontally in one dimension. Finally, for a process attribute in any given process, you may aggregate the ratings of all the process (attribute) outcomes for all the processes instances. This aggregation is performed as a matrix aggregation across the full scope of ratings (two dimensions).

Process attributes are rated using an ordinal scale. The assessor can choose to use his or her expert judgement to summarize the ratings without any formal mathematical approach, which is a typical approach to rating.

Alternatively, the assessor can use an aggregation method. This requires that the ordinal ratings be converted to interval values. For this to be valid two conditions must occur:

1. The scale must be sufficiently constrained to evenly spread the interval values. The rating scale shown below meets this requirement.
2. There must be evidence of an adequate sample size for adequate accuracy. Class 1 and Class 2 assessments are sufficiently rigorous to meet this need.

After meeting the above conditions then, the interval ratings can be converted to ordinal values as indicated below using either method:

N -> 0
P -> 1
L -> 2
F -> 3 or
N -> 0
P– -> 1
P±> 2
L– -> 3
L±> 4
F -> 5

The second scale allows for a finer degree of tolerance in the Partially and Largely ratings and thus is the recommended rating.

There are two one-dimensional methods available to summarize ratings. One dimension using an arithmetic mean and one dimension using median. Compute the average of the ratings of the interval values you rated earlier, rounding up or down to the nearest integer and then change that to the corresponding ordinal rating. When rounding, go down when the average value is less than the midpoint between consecutive integers, and round up when the average value is at or above the midpoint.

Another method is by computing the median values. Find the middle value of the ratings with the given data arranged in order from lowest to highest. If there is an odd number, the median is the middle value. If there is an even number of data, then take the average of the two middle values. If this results in a real number, then round the number to an integer value using the rules above. See Q & A Appendix A.

Sometimes, though, for a given process, the process outcomes of several process instances may need to be aggregated. This may be accomplished using expert judgement just as in a one-dimensional aggregation. Having converted the ordinal ratings to interval values, use one of the following two-dimensional aggregation methods.

Compute the arithmetic mean across the matrix of the full scope of ratings (expressed as numeric interval values) and then to the corresponding ordinal rating. Note, however, you can never take an average of averages as that would not be statistically valid. Another method would be a two-dimensional aggregation using heuristics where aggregation is performed using a defined set of rules to summarize as per Table 5.8 Process Capability Ratings. These aggregation methods were points that were not specified in ISO/IEC 15504.

KEY STEPS

Ensure that the defined set of assessment indicators are used
Select rating and aggregation methods
Verify traceability of evidence
Decide if the ordinal scale will be further refined for the measures P and L

Reporting

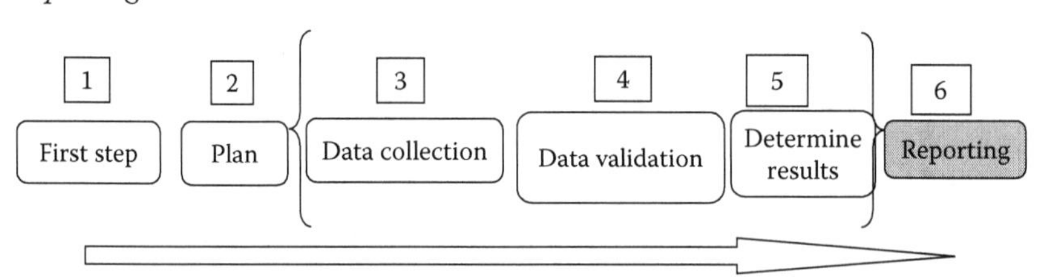

In this section, you compile the information collected that is relevant and supports understanding of the assessment. The results are presented in a way that offers comparison against other assessments, if required, as well as effective communication to the sponsor and other involved parties.

Your report should follow the table of contents shown below.

Example Report TOC

1. General
   a. Date of issue
   b. Version number
   c. Issuer name
   d. Distribution list
   e. Document change history

2. A summary of the assessment
3. The project assessed
   a. Identification of the PM and key staff
4. The class of assessment used
5. What type of assessment was done
6. What measurement process was used
   a. Identification of this process assessment model
7. The rating method and aggregation method used
8. The scope of the assessment
   a. The selected processes investigated
   b. Any exclusions and/or inclusions from/to the assessment scope
9. The projects selected as a basis for assessment, if applicable, detailing
   a. Project Name
   b. Short description
   c. Application area
   d. Project start date
   e. Project end date
   f. Current life cycle phase
   g. Project team size
   h. Project team person months.
10. The services selected as basis for assessment, and if applicable, the items below
    a. Identifier
    b. Short description;
    c. Application area;
    d. Service team size.
11. The assessment results
    a. Performance gaps and weaknesses established during process attribute rating
    b. Opportunities for improvement and risk mitigation, if applicable
    c. Any notes to the assessment
12. Any additional information collected during the assessment
13. The assessment report release
    a. Approved by the lead assessor
    b. Confirmed by all members of the assessment team

After you complete the report, the team should review it by for confirmation. It needs to be authorized by the assessment body. Finally, the Lead Assessor will approve it, and then the final report can be provided to the sponsor. Copies should be considered confidential and stored for later retrieval in a safe place.

KEY STEPS

Write report
Ensure recommended Table of Contents is followed

6

Executing the Assessment—Self-Assessment Guide

In this section, you will learn about the steps needed to perform a self-assessment. This type of assessment is used by organizations who wish to follow the assessment process but do not require or want to adhere to the more rigid demands of ensuring evidence and closely following all approvals.

NOTE

This section is designed to enable the reader to perform a self-assessment without necessarily referring to the earlier chapters. It contains all the steps you need to perform an assessment.

Self-Assessment Process

It is necessary to determine the need for an assessment of your project management (PM) processes. Typically, this is inspired by the realization that your projects, perhaps, are not performing at the level you desire, contributing to a decrease in business value. Alternatively, you may wish to assess the present capability of your PM processes. Regardless, this decision forms a portion of the start-up reason and needs to be ascertained to ensure you obtain the necessary results.

SELF-ASSESSMENT STEPS

Determine the business need
Establish roles and responsibilities
Assess competency requirements
Determine the assessment scope
Set up key interviews
Review collected data
Ensure that the defined set of assessment indicators are used
Write report

Remember, the basic purpose of process assessment is to understand and assess the processes implemented within your project management methodology. As this is a self-assessment, the class of assessment is not needed. You should proceed with the number of assessors and number of process instances you desire.

NOTE

The level of detail you collect, categories, classes, and how you perform ratings differ. They are either less formal or not required.

Each step is performed in order, repeating Steps 2 and 3 as necessary to complete all the process instances necessary for each Level. You may perform ratings as you validate the data rather than waiting until all data are collected for that particular level; thus, they are combined in the diagram shown in Figure 6.1.

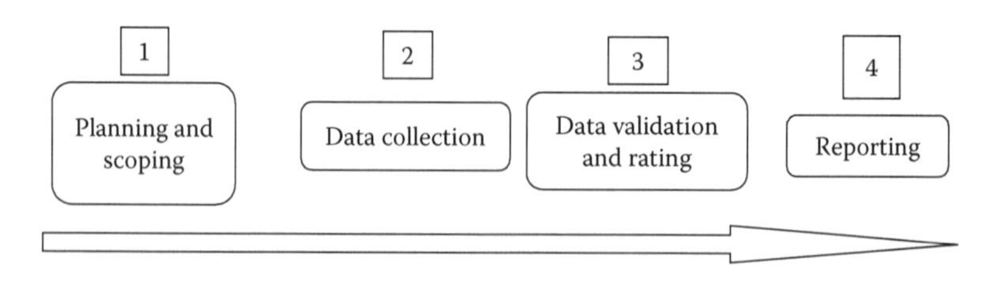

Figure 6.1 Assessment steps.

NOTE

Even though this is a self-assessment, having qualified people doing the work is a best practice to ensure an adequate level of competence.

Competency

If you want your self-assessment to have a higher level of assurance then you should ensure the team members are competent on the basis of appropriate education, training, and experience, including domain experience.

Scoping the Assessment

Your self-assessment might want to focus on the core planning processes. However, you might also determine the need by evaluating your projects and determining key areas of weakness.

In this key domain, there are 16 specific processes. You may want to prioritize which of these provide the biggest value as doing all sixteen processes may be too large an assessment. We suggest choosing one or two key processes (see Figure 6.2).

However, that is just one way to scope your assessment. You might also do a review of your previous projects and determine which domain provided the most difficulty in those projects and choose that for the assessment.

Assessing Each Attribute

Remember to begin your assessment by first assessing the Level 1 attribute, Process Performance. Here, you will use the individual process descriptions shown earlier for the particular Project Management processes you are including in your project scope. Determine whether the process outcomes are being achieved. To make this determination, review the base practices involved and

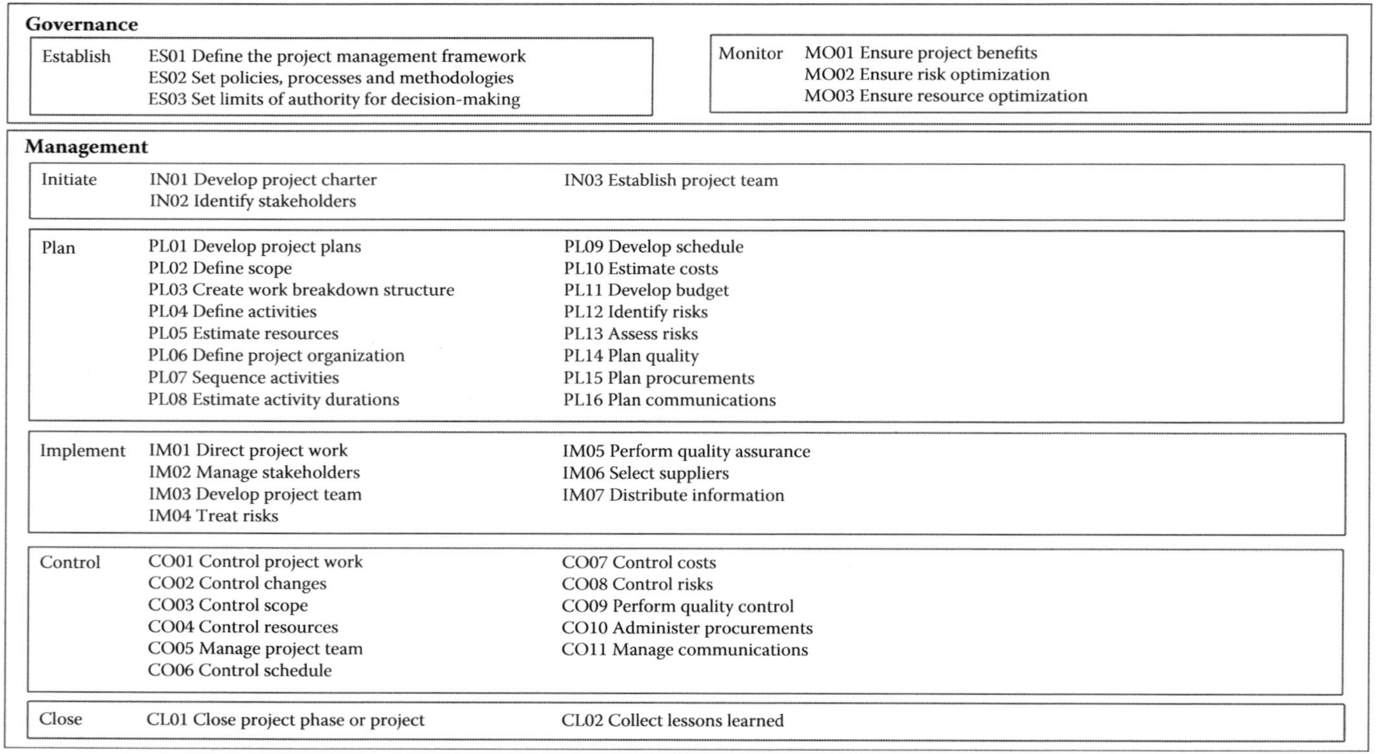

Governance

Establish		Monitor	
	ES01 Define the project management framework		MO01 Ensure project benefits
	ES02 Set policies, processes and methodologies		MO02 Ensure risk optimization
	ES03 Set limits of authority for decision-making		MO03 Ensure resource optimization

Management

Initiate	IN01 Develop project charter	IN03 Establish project team
	IN02 Identify stakeholders	
Plan	PL01 Develop project plans	PL09 Develop schedule
	PL02 Define scope	PL10 Estimate costs
	PL03 Create work breakdown structure	PL11 Develop budget
	PL04 Define activities	PL12 Identify risks
	PL05 Estimate resources	PL13 Assess risks
	PL06 Define project organization	PL14 Plan quality
	PL07 Sequence activities	PL15 Plan procurements
	PL08 Estimate activity durations	PL16 Plan communications
Implement	IM01 Direct project work	IM05 Perform quality assurance
	IM02 Manage stakeholders	IM06 Select suppliers
	IM03 Develop project team	IM07 Distribute information
	IM04 Treat risks	
Control	CO01 Control project work	CO07 Control costs
	CO02 Control changes	CO08 Control risks
	CO03 Control scope	CO09 Perform quality control
	CO04 Control resources	CO10 Administer procurements
	CO05 Manage project team	CO11 Manage communications
	CO06 Control schedule	
Close	CL01 Close project phase or project	CL02 Collect lessons learned

Figure 6.2 Process reference model.

assessing their level of compliance. To complete that review, you must ensure the appropriate work products are used as input and that the needed work product outputs are being created. This is the only time you refer to the processes as defined in the above process assessment model.

For higher level assessments, you use Generic Attributes, rather than the specific ones in Level 1. To achieve a level, the underlying level needs to be Largely or Fully complete. In Level 2 and above, you use generic attributes to assess each process. You can see in the diagram below that for Level 2 there are several generic practices and work products for you to review.

To indicate achievement of the attribute, all the generic practices in the second column along with the associated generic work products in third column need to be achieved for the particular process you are assessing. As you can see, rather than the specifics of Level 1, which apply directly to the process you are assessing, these same generic items apply to any of the Project Management processes you choose to assess.

Example 6.1: PA 2.1 Performance Management

This process attribute is a measure of the extent to which the performance of the process is managed. Full achievement of this attribute results in the process achieving the following attributes (see Table 6.1).

Example 6.2: PA 2.2 Work Product Management

This process attribute is a measure of the extent to which the work products produced by the process are appropriately managed. Full achievement of this attribute results in the process achieving the following attributes (see Table 6.2).

Remember, there is no particular need to be at any given Level other than Level 1. After all, why have processes if they do not actually perform as desired? To go higher than that is a business decision based on your need and desire to expend the additional effort needed to maintain a process at each successive higher level.

Table 6.1 PA 2.1 Performance Management

RESULT OF FULL ACHIEVEMENT OF THE ATTRIBUTE	GENERIC PRACTICES (GPs)	GENERIC WORK PRODUCTS (GWPs)
a. Objectives for the performance of the process are identified.	*GP 2.1.1 Identify the objectives for the performance of the process.* The performance objectives, scoped together with assumptions and constraints, are defined and communicated.	*GWP 1.0 Process documentation* should outline the process scope. *GWP 2.0 Process plan* should provide details of the process performance objectives.
b. Performance of the process is planned.	*GP 2.1.2 Plan the performance of the process to fulfill the identified objectives.* Basic measures of process performance linked to business objectives are established. They include key milestones, required activities, estimates and schedules.	*GWP 2.0 Process Plan* should provide details of the process performance objectives. *GWP 9.0 Process performance records* should provide details of the outcomes.
c. Performance of the process is monitored.	*GP 2.1.3 Monitor the performance of the process to fulfill the identified objectives.* Basic measures of process performance linked to business objectives are established and monitored. They include key milestones, required activities, estimates, and schedules.	*GWP 2.0 Process Plan* should provide details of the process performance objectives. *GWP 9.0 Process performance records* should provide details of the outcomes.
d. Performance of the process is adjusted to meet plans.	*GP 2.1.4 Adjust the performance of the process.* Action is taken when planned performance is not achieved. Actions include identification of process performance issues and adjustment of plans and schedules as appropriate.	*GWP 4.0 Quality record* should provide details of action taken when performance is not achieved.
e. Responsibilities and authorities for performing the process are defined, assigned and communicated.	*GP 2.1.5 Define responsibilities and authorities for performing the process.* The key responsibilities and authorities for performing the key activities of the process are defined, assigned and communicated. The need for process performance experience, knowledge, and skills is defined.	*GWP 1.0 Process documentation* should provide details of the process owner and who is responsible, accountable, consulted, and informed (RACI). *GWP 2.0 Process plan* should include details of the process communication plan as well as process performance experience, skills, and requirement.

(*Continued*)

Table 6.1 (*Continued*) PA 2.1 Performance Management

RESULT OF FULL ACHIEVEMENT OF THE ATTRIBUTE	GENERIC PRACTICES (GPs)	GENERIC WORK PRODUCTS (GWPs)
f. Personnel performing the process are prepared for executing their responsibilities.	*GP 2.1.6 Assign personnel and train them to perform the process according to plan.* Personnel performing processes accept responsibility and have the necessary skills and experience to carry out their responsibilities.	*GWP 2.0 Process plan* should include details of the process communication plan as well as process performance experience, skills and requirement.
g. Resources and information necessary for performing the process are identified made available, allocated and used.	*GP 2.1.7 Identify and make available resources to perform the process according to plan.* Resources and information necessary for performing the key activities of the process are identified, made available, allocated, and used.	*GWP 2.0 Process plan* should provide details of the process training plan and process resourcing plan.
h. Interfaces between the involved parties are managed to ensure effective communication and clear assignment of responsibility.	*GP 2.1.8 Manage the interfaces between involve parties.* The individuals and groups involved with the process are identified, responsibilities are defined, and effective communication mechanisms are in place.	*GWP 1.0 Process documentation* should provide details of the individuals and groups involved (suppliers, customers, and RACI). *GWP 2.0 Process plan* should provide details of the process communication plan.

Table 6.2 PA 2.2 Work Product Management

RESULT OF FULL ACHIEVEMENT OF THE ATTRIBUTE	GENERIC PRACTICES (GPs)	GENERIC WORK PRODUCTS (GWPs)
a. Requirements for the work products of the process are defined.	*GP 2.2.1 Define the requirements for the work products.* This should include content structure and quality criteria.	*GWP 3.0 Quality plan* should provide details of quality criteria and work product content and structure.
b. Requirements for documentation and control of the work products are defined.	*GP 2.2.2 Define the requirements for documentation and control of the work products.* This should include identification of dependencies, approvals and traceability of requirements.	*GWP 1.0 Process documentation* should provide details of controls; for example, control matrix. *GWP 3.0 Quality plan* should provide details of work product, quality criteria, documentation requirements, and change control.
c. Work products are appropriately identified, documented, and controlled.	*GP 2.2.3 Identify, document and control the work products.* Work products are subject to change control, versioning, and configuration management as appropriate.	*GWP 3.0 Quality plan* should provide details of work product, quality criteria, documentation requirements, and change control.
d. Work products are reviewed in accordance with planned arrangements and adjusted as necessary to meet requirements.	*GP 2.2.4 Review and adjust work products to meet the defined requirements.* Work products are subject to review against requirements in accordance with planned arrangements and any issues arising are resolved.	*GWP 4. Quality records* should provide and audit trail of reviews undertaken.

When collecting data, the data needs to be sufficient to provide coverage of the organization and process scope for the assessment. You will want to collect the data using direct or indirect evidence that satisfies your assessment criteria. Direct evidence would be written project plans, the project charter, the specific project management framework in use, and so on, indirect evidence would be interview notes and discussions with project participants.

NOTE

Remember, with a self-assessment, unlike the earlier formal assessment process, there is no requirement to collect evidence. You can collect the amount of evidence you want in order to remain confident in your assessment.

It is recommended that you use the following information as a guideline for managing any evidence you collect.

Number each domain and the corresponding processes with a Process ID. Within each process in that domain, each outcome, base practice, and work product are then also numbered. Use these numbers to reference any materials collected, whether interview notes or documents with these numbers and a further number identifying the actual piece of evidence.

See Appendix H: Sample Data Tracking Form for an example of how you might track materials supporting your assessment.

Make sure the document or interview is worthwhile and accurate relevant to the information you need. Ensure that you collect sufficient data to be able to readily validate its applicability and ability to determine if it is objective enough for your needs. Once collected, rate the information using the scale shown in Table 6.3.

Table 6.4 outlines the attributes that must be achieved within each Level for you to be able to indicate you have achieved them. If you desire to indicate your processes have achieved a particular rating, the information in this table will assist you in achieving that goal.

Table 6.3 Rating Scale and Percentage

RATING	MEANING	LEVEL OF ACHIEVEMENT
N	Not achieved	0%–≤15%
P–	Partially achieved minus	>15%–≤32.5%
P+	Partially achieved plus	>32.5%–≤50%
L–	Largely achieved minus	>50%–≤67.5%
L+	Largely achieved plus	>67.5%–≤85%
F	Fully achieved	>85%–≤100%

Table 6.4 Process Capability Ratings

SCALE	PROCESS ATTRIBUTES	RATING
Level 1	Process Performance	Largely or fully
Level 2	Process Performance	Fully
	Performance Management	Largely or fully
	Work Product Management	Largely or fully
Level 3	Process Performance	Fully
	Performance Management	Fully
	Work Product Management	Fully
	Process Definition	Largely or fully
	Process Deployment	Largely or fully
Level 4	Process Performance	Fully
	Performance Management	Fully
	Work Product Management	Fully
	Process Definition	Fully
	Process Deployment	Fully
	Quantitative Analysis	Largely or fully
	Quantitative Control	Largely or fully
Level 5	Process Performance	Fully
	Performance Management	Fully
	Work Product Management	Fully
	Process Definition	Fully
	Process Deployment	Fully
	Quantitative Analysis	Fully
	Quantitative Control	Fully
	Process Innovation	Largely or fully
	Process Innovation Implementation	Largely or fully

Source: Permission to use extracts from ISO/IEC 33063:2015 was provided by the Standards Council of Canada (SCC). No further reproduction is permitted without prior written approval from SCC.

Reporting

Once your assessment reaches this stage it is time to document your results. You can write a report or use a summary table such as the one shown below:

Process self-assessment report

Domain	Process ID	Name	Assessed	Level 0	Level 1	Level 2	Level 3	Level 4	Level 5
Establish									
	ES01	Define the project management framework							
	ES02	Set policies, processes, and methodologies							
	ES03	Set limits of authority for decision-making	✓		F	P+			
Monitor									
	MO01	Ensure project benefits							
	MO02	Ensure risk optimization							
	MO03	Ensure resource optimization	✓		F	L			
Initiate									
	IN01	Develop project charter							
	IN02	Identify stakeholders							
	IN03	Establish project team							
Plan									

Rating
N– 0%–15% Not achieved
P– 15% ≤ 32.5% Partially minus
P+ 32.5% ≤ 50% Partially plus
L– 50% ≤ 67.5% Largely minus
L+ 67.5% ≤ 85% Largely plus
F– 85%–100%

In this section, we have attempted to provide the information needed to allow you to perform a capability assessment on your selected project management processes. Using this section, you can easily determine the strengths and weaknesses involved and determine your path forward to strengthening management of your projects.

Appendix A: Level 1 Output Work Products

Following are descriptions of the documents or artifacts and minimum requirements for the process reference model mentioned in Part 1.

WP ID	WP	DESCRIPTION
ES01-WP1	Project management guidelines	Document outlining a standard approach for project management. It should include: • Detailed project management methodology • Tailoring instructions
ES01-WP2	Project charter	Document giving an individual authority to act as project manager for the project. It provides the mandate to run the project and it turns the project from an idea into an actual program of work, with allocated owners (and agreement on funding). It should include: • Introduction and purpose • Project overview • Justification • Business need • Business impact • Strategic alignment • Scope

(*Continued*)

WP ID	WP	DESCRIPTION
ES01-WP2 (Continued)	Project charter (Continued)	• High-level requirements • Deliverables • Boundaries • Duration • Timelines • Milestones • Budget • Funding source • Estimates • Alternative Analysis • Assumptions • Constraints • Risks • Project organization • Roles and responsibilities of the project manager to assign resources to the team • Stakeholders • Project charter approval
ES02-WP1	Project management policy	Document that provides direction on appropriate systems, processes and controls for managing projects are in place and support the achievement of project outcomes while limiting the risk to stakeholders. It should include: • Purpose • Scope • Policy statements • Responsibility • Definitions • Associated documents • Authority • Review • Sanctions • Effective date
ES02-WP2	Project management methodology	A document that provides a "how-to" for governing, managing and delivering successful projects. It should include: • Introduction • Project governance • Project delivery framework • Project processes • Project procedures • Project deliverables

(*Continued*)

WP ID	WP	DESCRIPTION
ES02-WP3	Project management process definition documents	A document that describes the various processes of the project management methodology. It should include: • Process description • Process purpose • Process outcomes • Best practices • Inputs • Outputs
ES03-WP1	Authority limits	A document that provides the limits of the project and project manager's authority to control project variation. It should include: • Assignment of accountabilities, responsibilities and authorities • Exception management • Change budget • Risk budget or contingency
ES03-WP2	Risk tolerance	Approved risk tolerance levels from a standard risk management process. It should include: • Cost tolerance • Schedule tolerance • Quality tolerance • Product or deliverable specification • Benefit tolerance
MO01-WP1	Business case document	A document to justify the undertaking of a project based on the estimated costs against the anticipated benefits derived from the project offset by any risk. The business case is refined and monitored as the project progresses to ensure the on-going viability of the project. It should include: • Executive summary • Purpose • Business options • Expected benefits • Expected opportunity costs • Timescale • Costs • Investment appraisal • Major risks

(*Continued*)

WP ID	WP	DESCRIPTION
M001-WP2	Benefits review plan	Defines how and when a measurement of the achievement of the project's benefits should be made. It should include: • Scope of the benefits review plan covering what is measured • Documents who is accountable for the expected benefits • Methods to measure achievement of expected benefits • Resources needed to perform review work • Baseline measures used to calculate improvements
M002-WP1	Risk management policy	Policy that outlines the enterprise risk management requirements, roles, responsibilities and controls. It should include: • Purpose • Scope • Policy statements • Responsibility • Definitions • Associated documents • Authority • Review • Sanctions • Effective date
M002-WP2	Risk appetite	A document that describes the level of acceptable risk. It should include: • Objectives • Risk attitude • Risk culture • Risk perception • Risk factors such as risk propensity and risk preferences • Risk thresholds • Risk capacity • Risk exposure • Risk actions such as amount and type of risk that the organization will assume

(*Continued*)

WP ID	WP	DESCRIPTION
MO03-WP1	Resource availability	Document showing the aggregate of the resource's accessibility, reliability, maintainability, serviceability, and ability to be secured. It should include: • *Resource availability*: renewable resources • *Resource demand*: activity resource requirements • *Resource conflicts*
IN01-WP1	Project charter	Define the project to form the basis for its management and overall assessment. Provides direction for and scope of the project and along with the project plan its forms the "contract" between the governing body and the managing body, also known as the project manager. It should include: • Project definition • Project approach • Business case • Project management team structure • Role descriptions • Quality management strategy • Risk management strategy • Change management strategy • Communications management strategy • Project plan • Project controls • Summary of process tailoring
IN02-WP1	Stakeholder register	A document that contains the information about the project's stakeholders and identifies the people, groups and organizations that have any kind of interest in the project. It should include: • Different stakeholders • Type of stakeholder • Role in project • Level of interest • Level of influence on project • Concerns • Requirements • Expectations • Type of communication

(*Continued*)

WP ID	WP	DESCRIPTION
IN03-WP1, IM03-WP1	Team performance	Document used to adjust the composition, context or direction of a team to increase the effectiveness of the team through measurement. It should include: • Organizational benchmarks for teams • Comparison with expected progress or outcomes of the team's work
IN03-WP2, IM03-WP2	Team appraisal	Document that provides feedback to the project team. It should include: • Management's expectations • Defined roles • Communication • Decision-making • Performance evaluation • Customer satisfaction • Product or deliverable backlog
PL01-WP1	Project plans	Statement of how and when objectives are achieved by showing the major deliverables, activities and resources. It should include: • Plan description • Plan prerequisites • External dependencies • Planning assumptions • Lessons incorporated • Monitoring and control • Budgets • Tolerances • Product or deliverable description • Schedule
PL01-WP2	Project management plan	The plan of plans used to guide both project execution and project control. It should include: • Scope plan • Requirements plan • Schedule plan • Financial plan • Quality plan • Resource plan • Stakeholder plan • Communication plan • Change plan • Staffing plan • Risk plan • Benefits plan

(*Continued*)

WP ID	WP	DESCRIPTION
PL02-WP1	Scope statement	A document that outlines the project's deliverables, results and outcomes, and identifies the constraints, assumptions and key success factors. It should include: • Project charter • Project stakeholders • Scope description: in scope and out of scope • High level project requirements • Project goals and objectives • Project boundaries • Project strategy • Project deliverables • Acceptance criteria • Project constraints • Project Assumptions • Milestones • Cost estimates • Cost benefit analysis
PL02-WP2	Requirements	A documentation of the conditions or tasks that must be completed to ensure the success or completion of the project. It should include: • Project • Date(s) • Preparer or source • Document status • Scope • Business case • Product or deliverable requirements • Functions • Use case • User • System • Interface • Assumptions • Constraints
PL03-WP1	Work breakdown structure (WBS)	Package about one or more required products or deliverables collated by the project manager to pass responsibility for work or delivery formally to a team member. It should include: • Title • Date • Team manager or authorized person

(*Continued*)

WP ID	WP	DESCRIPTION
PL03-WP1 (Continued)	Work breakdown structure (WBS) (Continued)	• WBS description • Techniques, tools and procedures • Development interfaces • Operations and maintenance interfaces • Change management requirements • Joint agreements • Tolerances • Constraints • Reporting arrangements • Problem handling and escalation procedure • Extract or references especially project plan and product or deliverable description • Approval method for WBS completion
PL03-WP2	Work breakdown structure dictionary	A document that provides a detailed information about each element in the WBS, including work packages and control accounts. It should include: • Work package ID • Work package name • Work package description • Assignee • Date assigned • Date due • Estimated cost
PL04-WP1	Activity list	An itemized documentation of all of the schedule activities that are part of a particular project. It should include: • WBS reference • Activity name • Assignee • Project manager • Due date • Description • Priority • Predecessor activity • Cost • Baseline • Actual • Activity detail • Duration • Resource
PL05-WP1	Resource requirements	A document specifying the resources required for the project. It should include: • Resource type • Estimated requirements • Availability

(*Continued*)

WP ID	WP	DESCRIPTION
PL05-WP2	Resource plan	A document used throughout the project to plan for needed resources. It should include: • Staffing • Incremental staffing • Equipment • Other expenses • Schedule • Capacity • Capability
PL06-WP1	Role descriptions	A document that defines roles and responsibilities for the project. It should include: • Role • Responsibilities • Duties
PL06-WP2	Project organization chart	A diagram that shows the hierarchical structure of an organization. It should include: • All identified roles • Relationships • Reporting structure
PL07-WP1	Activity sequence	A document that shows logical sequence of work to obtain the greatest efficiency given all project constraints by identifying and documenting relationships among the project activities. It should include: • WBS activities • Sequence • Type of relationship • A finish-to-start relationship • A start-to-start relationship • A finish-to-finish relationship • A start-to-finish relationship • Task start dates • Task end dates • Lag and lead times • Critical path • Slack
PL08-WP1	Activity duration estimates	A documentation of activities and estimates of their duration using an estimation technique. It should include: • Activity list • Activity attributes • Scope statement

(*Continued*)

WP ID	WP	DESCRIPTION
PL08-WP1 (Continued)	Activity duration estimates (Continued)	• Lessons learned • Level of effort • Duration • Resource requirements • Resource calendars • Assumptions • Constraints
PL09-WP1	Schedule	A listing of a project's milestones, activities, and deliverables, usually with intended start and finish dates. It should include: • WBS • Resource • Start date • End date • Duration • Milestones
PL10-WP1	Cost estimates	A document that is a product of the cost estimating process. It should include: • Activity • Duration • Resource • Standard resource cost • Type of estimate • Estimate classification • Estimate quality • Contingency
PL11-WP1	Budget	A document that shows the funds authorized to execute the project. It should include: • Cost estimates
PL12-WP1	Risk register	A document to record details of, monitor and manage the project's risks. It should include: • Risk ID • Risk description • Risk owner • Date opened • Target date • Date closed • Probability • Impact • Contingency plan • Current status

(*Continued*)

WP ID	WP	DESCRIPTION
PL13-WP1	Prioritized risks	A list of project risks prioritized by probability and impact. It should include: • Priority and impact ranking technique • Risk list • Probabilities • Impacts • Priorities
PL14-WP1	Quality plan	Document defining the acceptable level of quality as defined by the customer, and describing how the project will ensure this level of quality in its deliverables and work processes. It should include: • Quality objectives • Key project deliverables and processes to be reviewed for satisfactory quality level • Quality standards • Quality control and assurance activities • Quality roles and responsibilities • Quality tools • Plan for reporting quality control and assurance problems
PL15-WP1	Procurement plan	A document that defines the products and services that you will obtain from external suppliers. It should include: • Procurement Management Approach • Contract Types • Procurement Constraints • Scope • Responsibilities and Authorities • Decision Criteria • Procurement Documentation • Procurement Risk • Reporting Formats • Supplier Performance Measurement
PL15-WP2	Preferred suppliers list	A document that contains a list of approved suppliers that the organization has assessed and continuously follow up with different KPIs. It should include: • Preferred vendors • Products or services offered • Contact information
PL15-WP3	Make-or-buy decision list	A list of all components and whether the project will make or produce internally or buy it from the market. It should include: • Alternatives • Costs • Benefits

(*Continued*)

WP ID	WP	DESCRIPTION
PL16-WP1	Communications plan	A documentation of the modes of messaging to a project's affected stakeholders. It should include: • Communication objectives • Target audiences • Key content for the communications • Communication method and frequency
IM01-WP1	Progress data	Data that shows the progress and status of a project, crucial to the effective management of a project. It should include: • Status • Completion date forecast • Variance analysis • Earned value
IM01-WP2, IM02-WP2	Issues log	A log containing the description, impact assessment and recommendations for change, off-specification, and a problem or a concern. It should include: • Issue Identifier • Issue type • Date raised • Raised by • Issue reporter • Issue description • Impact analysis • Recommendation • Priority • Severity • Decision • Approved by • Decision date • Closure date
IM01-WP3, CL02-WP1	Lessons learned	Document to pass on any lessons from the phase or project that could be used in another phase or project. It should include: • Executive summary • Scope of the report; that is, phase or project • Review of what went well, what went badly and any recommendations for corporate or PMO consideration. This may include as appropriate: • Processes used or not used

(*Continued*)

WP ID	WP	DESCRIPTION
IM01-WP3, CL02-WP1 (Continued)	Lessons learned (Continued)	• Any specialist methods used, such as EVM • Project strategies for risk, quality, communications, time and cost management • Project controls • Abnormal events causing deviations from plan • Review of useful measurements
PL15-WP4, IM02-WP1, IM04-WP2, IM05-WP1, CO01-WP1, CO03-WP1, IM02-WP1, CO04-WP3, CO06-WP1, CO07-WP3, CO08-WP1, CO09-WP4, CO10-WP1	Change requests	A document completed for each major change to a project's scope, budget or schedule. It should include: • Date • Priority or urgency • Definition of the problem or opportunity • Benefits • Costs • Alternate solutions • Decision rationale • Project impact • Recommendations • Approvals and sign-offs
IM04-WP1	Risk responses	A document that shows options and actions to enhance opportunities, and to reduce threats to project objectives. It should include: • Risks • Risk strategies • Residual risk • Secondary risk
IM06-WP1	Request for information, proposal, bid, offer or quotation	A document that requests additional information, a proposal or quotation for a product or service. It should include: • Problem statement • Type of solicitation • Rules of engagement • Dates • Product or service requirements • Selection criteria
IM06-WP2	Contracts or purchase orders	An agreement between two or more parties to accomplish a certain goal in a prescribed manner underlying the project. It should include: • Dates • Type of contract • Statement of work

(*Continued*)

WP ID	WP	DESCRIPTION
IM06-WP2 (Continued)	Contracts or purchase orders (Continued)	• Engagement and performance • Terms • Conditions • Security requirements • Standard contract provisions
IM06-WP3	Selected suppliers list	A document that contains a list of selected suppliers. It should include: • Selected vendors • Products or services offered • Contact information
IM07-WP1	Distributed information	Documentation that demonstrates that the project shares and the stakeholders receive the necessary information. It should include: • Recipient • Information • Date
CO01-WP2	Progress reports	A documentation of phase or project progress. It should include: • Summary • Objectives • Progress vis a vis plans • % age completed • Work-in-progress • Remaining work • Problems or issues • Changes • Tasks
CO01-WP3	Project completion reports	A formal document closing a project. It should include: • Summary • Overview • Highlights • Objectives • Results and outcomes • Financial statement • Lessons learnt
CO02-WP1	Approved changes	An approved change is a change request that has been submitted by the requestors, reviewed by the appropriate parties through use of the integrated change control process, and granted authorization to be take place. It should include: • Requestor • Reviewers • Approver

(*Continued*)

WP ID	WP	DESCRIPTION
C002-WP2	Change register	A list of the information relating to changes in a project. It should include: • Nature of the change • Impact of the change • Change approval details, and status of the approval • Change implementation schedule and date • Current status of all changes
C004-WP4, C006-WP2, C007-WP4, C008-WP2, C009-WP5, C010-WP2, C011-WP2	Corrective actions	A report containing the description, impact assessment, and recommendations for a request for change. It should include: • Identifier • Date raised • Raised by • Report author • Description of corrective action • Impact analysis • Recommendation • Priority • Severity • Decision • Approved by • Decision date • Closure date
C005-WP1	Staff performance	Document used to increase the effectiveness of a team member through measurement. It should include: • Organizational benchmarks for individuals • Comparison with expected progress or outcomes of the individual's work
C005-WP2	Staff appraisals	Document that provides feedback to the project staff. It should include: • Management's expectations • Defined role • Performance evaluation
C007-WP1	Actual costs	A document that represents the true total and final costs accrued for a period allocated for all scheduled activities. It should include: • Actual start date • Actual end date • Actual costs of work performed (ACWP) • Itemized costs • Direct labour • Indirect labour • Materials • Variance analysis

(*Continued*)

WP ID	WP	DESCRIPTION
C007-WP2	Forecasted costs	A document used to control the cost performance of a project and to predict the final project cost. It should include: • Current actual costs (AC) • Planned Cost of Work Remaining (PCWR) • Expected at Completion (EAC)
C009-WP1	Quality control measurements	A document used to analyze as well as evaluate the quality of the different processes involved in the project against organizational standards or on the specified project requirements. It should include: • Measurement method • Standard or requirement • Actual measurement • Verification • Validation
C009-WP2	Verified deliverables	A list of output from project tasks that meet quality control measures as specified in Quality Management Plan. It should include: • Requirement or deliverable • Quality criteria • Quality control measures
C009-WP3	Inspection reports	A report showing that the deliverable or product was inspected for compliance. It should include: • Inspector • Inspection criteria or specification • Description • Compliance • Corrective actions
C011-WP1	Accurate and timely information	All information should meet the needs of the user. It should include: • Information criteria • Information review
CL01-WP1	Completed procurements	A list of completed procurements. It should include: • Requirement • Contract • Award criteria • Procurement status • Validated deliverables • Vendor metrics • Payment status

(*Continued*)

WP ID	WP	DESCRIPTION
CL01-WP2	Project or phase closure report	Reviews phase or project closure for how well the project performed against the version of project initiation documentation used to authorize it. It should include: • Project manager's report summarizing performance • Review of the business case • Review of project or phase objectives • Review of team performance • Review of products including quality activities and records, approval records, off-specifications, product handover, and summary of follow-on action recommendations • Issues and risks (Phase report) • Forecast (Phase report)
CL01-WP3	Released resources	A checklist for deciding when to release resources when no longer required to meet the objectives of the project. It should include: • Work package summary • Validated deliverables or products • Performance review • Performance appraisal • Contract review

Appendix B: Level 2–5 Generic Work Products

In this appendix, you will find a description of all the generic work products (GWP) mentioned in the book. You need these work products or artefacts to assess Levels 2–5.

GWP ID	GWP	EXPECTED CONTENTS	RELATED GP	DESCRIPTION
1.0	Process documentation	Process name	–	The name of the process
		Process owner	GP 2.1.4	The person responsible for the process
		Process scope	GP 2.1.1	A clear statement of where the process starts and ends
		Process roles	GP 2.1.6	Details of key role players such as suppliers (inputs) and customers (outputs)
		Process map	GP 3.1.2	A schematic of the process sequential flow in a swim lanes or LOVEM diagram/flow chart
		RACI chart	GP 2.1.4 GP 2.1.6	Identifies responsible, accountable, consulted, and informed roles for each key activity

(*Continued*)

GWP ID	GWP	EXPECTED CONTENTS	RELATED GP	DESCRIPTION
1.0 (Continued)	Process documentation (Continued)	Internal control matrix	GP 2.2.2	Shows identified risk for the key activities
		Process procedures	GP 3.1.1	Document outlining the activities required to achieve the process outcomes
2.0	Process plan	Process performance objectives	GP 2.1.1 GP 2.1.2	Each process plan will, minimally, have targets such as milestones/inch pebbles, activities, estimated input and output volumes, or schedules
		Process resourcing	GP 2.1.5 GP 3.2.4	A plan indicating resources, such as people, and information required to meet the performance required for the process, and information on what resources are to be supplied
		Process communication	GP 2.1.4 GP 2.1.6 GP 3.2.3	A communication plan for the process. It should minimally include the following: a. RACI for communication b. Target stakeholders c. Content to communicate d. Timing for communications e. Communication media and type
		Process infrastructure and work environment	GP 3.1.4 GP 3.2.5	Facilities, tools, methods and work environment for performing the processes. This is found in ISO 21500, PMBOK and PRINCE2 among others
		Process performance experience, skills requirement	GP 2.1.4	Job descriptions and skills required to undertake the process
		Process training requirement	GP 2.1.5	User skills and competencies, including individual training requirements

(*Continued*)

GWP ID	GWP	EXPECTED CONTENTS	RELATED GP	DESCRIPTION
3.0	Quality plan	Statement of quality policy and objectives	GP 2.1.2	A statement of the customer's quality expectations—Critical to Quality—for the process; for example, deliverables, dates and costs
		Work products content	GP 2.2.1	Identification of all work products, their structure or format, and expected content
		Quality criteria for the work products produced by the processes	GP 2.2.1	The criteria against which each work product will be reviewed and approved
		Work products documentation	GP 2.2.2	Requirements for documentation and control requirements, such as identification, traceability, and approvals
		Work products change control, versioning and configuration management requirements	GP 2.2.3	Outline of procedures for versioning and change control to be applied to work products
4.0	Quality records	Records of reviews against requirements and action taken	GP 2.2.4	Record of reviews undertaken of work products together with any issues arising and their resolution
5.0	Policies and standards	Operational objectives and responsibility for the process	GP 3.1.1	Statement of the organization's objectives for the process as it is applied across organizational units. It should identify overall responsibility for the process
		Minimum standard of performance required for a process	GP 3.1.1	The expected level of performance expected for the process across the organization. This might include milestones/inch pebbles, required activities, estimated output volumes, and schedules

(*Continued*)

GWP ID	GWP	EXPECTED CONTENTS	RELATED GP	DESCRIPTION
5.0 (Continued)	Policies and standards (Continued)	Standard process mapping, including expected sequence and interaction between processes	GP 3.1.2 GP 3.2.1	Schematic picture, such as flowchart or "swim lanes" chart, showing sequential flow of work expected for the process. This should also identify expected interactions between different implementations of the process
		Standardized procedures	GP 3.2.1	Document providing the procedures that should be followed in all implementations of the processes
		Roles and competency for performing the process to minimum standards of performance	GP 3.1.3 GP 3.2.2 GP 3.2.3	Standardized job descriptions, experience, qualifications, and skills requirements for the process
		The minimum infrastructure, such as facilities, tools and methods, and work environment for performing the standard process	GP 3.1.4	The facilities, tools, methods, and work environment for performing the processes
		Reporting and monitoring requirements, including audit and review	GP 3.1.5	Expected reports and monitoring required for the process, including standardized reporting requirements
6.0	Performance improvement plan	Process improvement objectives	GP 5.1.1	The level of performance expected from the process, based on business objectives
		Analysis against best practice	GP 5.1.3	Identified opportunities for process improvements based on analysis of comparison with industry best practices
		Technology improvement opportunities	GP 5.1.4	Identified opportunities for process improvements based on analysis of technology and process innovations

(*Continued*)

GWP ID	GWP	EXPECTED CONTENTS	RELATED GP	DESCRIPTION
6.0 (Continued)	Performance improvement plan (Continued)	Improvement actions	GP 5.1.5	Identified actions for improving the process across the organization
		Improvement implementation plan	GP 5.1.6	The proposed improvements, planned actions to implement those improvements, responsibilities, and timetable
		Project quality approach	GP 5.1.5	Proposed process for confirming the achievement of the improvements; for example, measures and reviews
7.0	Process measurement plan	Measurement objectives	GP 4.1.1	Quantitative objectives for the process relative to quality and process performance, based on customer needs and business objectives
		Proposed measures/ indicators	GP 4.1.2	Identification of what is to be measure and the measurement indicators
		Data collection procedures	GP 4.1.3	Identification of how data is collected to support measurement
		Analytical procedures	GP 4.1.3 GP 4.1.4	Identification of analytical procedures to be used, from simple charts and graphs, to more sophisticated quantitative analyses, such as statistical process control (SPC), structural modelling, or other multivariate statistical methods
8.0	Process control plan	Control techniques	GP 4.2.1	Description of the methods used to minimize process and product variation. It will differ for each process and may include standards, testing, reviews, and walkthroughs
		Measurement approach	GP 4.2.1	How the variation in each process is measured
		Control limits for normal performance	GP 4.2.2	Process specification or the acceptable level of process variation
9.0	Process performance records	Records of reviews against requirements and action taken	GP 4.1.5	Record of actual process performance, with any variations from expected results and action taken to correct variations

Appendix C: Frequently Asked Questions (FAQ)

Q1. What is a pain point?

A1. A pain point is something that keeps coming back like a bad burrito or showing up like a bad penny. Pain points are often unpleasant or unwanted events, especially ones that repeatedly appear at inopportune times. You keep getting called out by Audit for the same issue or problem on each and every audit; that is a pain point. Some project management pain points include:

- Business refuses to work with the project management office
- Projects are repeatedly over budget or delivered late
- Projects fail to meet business objectives and do not deliver value
- Failure to meet the requirements in the project charter
- Projects are not delivering benefits
- High project team turnover
- Project team member burnout
- Complex project management methodology

Q2. How do I use pain points to pick processes?

A2. Use pain points as input to the process improvement process and plan. For example, if stakeholder participation is low, you have calculated the stakeholder involvement index, but it is very disappointing. Then you would use that information to focus on processes IN02, PL13, PL16, IM02, IM07, CO02, CO11, and CL01 and most importantly processes IN02: Identify stakeholders and IM02: Manage stakeholders.

Q3. How do I align project management goals with business goals?
A3. An important part of any assessment is deciding on the scope of the assessment. Your organization most likely could not absorb the changes from more than a couple of processes and most likely would not tolerate a never-ending assessment. Thus, you need to focus your assessment on those items that really matter and what matters are those processes that contribute strongly to meeting organizational objectives. Consequently, pick those processes that align with them. This act lies on a continuum from easy to difficult. In the area of enterprise information technology projects, you could leverage the COBIT 5 cascading goals mechanism to help with alignment. Using COBIT 5, you map Stakeholder Needs → Enterprise Goals → IT-related Goals → Process Goals. For instance, we could do the following mapping: Do IT projects fail to deliver what they promise—and if so, why? (Stakeholder need) → Stakeholder value of business investments (Enterprise Goal) → Managed IT-related business risk (IT-related Goal) → Ensure benefits delivery (Process Goal).

Or perhaps your organization uses a Balanced Scorecard[1] and you have done strategy mapping.[2] Should you not know about strategy mapping, it is a visual tool for showing how strategies align with and support each other. It is comparable in a way to a goals-cascading mechanism. Simplistically, you map Customer value proposition → Mission statement → Core values → Strategic vision → Overall goal → Processes. You can see these two methods align well. So, we recommend you use COBIT 5 and ISO 38500[3] when working on IT projects and fudge your alignment by using COBIT 5 when working on non-IT projects.

---

[1] Refer to *The Balanced Scorecard: Translating Strategy into Action* by Robert S. Kaplan and David P. Norton.

[2] Refer to *Strategy Maps: Converting Intangible Assets into Tangible Outcomes* by Robert S. Kaplan and David P. Norton.

[3] Refer to ISO/IEC 38500:2015, *Information technology – Governance of IT for the organization.*

Lastly, you could use the OPM3 Self-Assessment Questions in Appendix X3 of the book to determine what processes to consider. Appendix I provides a way to rank processes for selection.

Q4. But every project is unique so how could I assess project management?

A4. By definition, projects are unique. As Heraclitus once said, "No man ever steps in the same river twice, for it's not the same river and he's not the same man." So, although projects are similar, they are indeed unique: each one has its own requirements and resources. However, then you are not evaluating a particular project, but the processes used to deliver the project. Project management, as defined in Section 3.3 of ISO 21500, is "the application of methods, tools, techniques and competencies to a project." Whether the project manager chooses to use a specific process or tool is within their purview. We are not directly assessing this part, nor does ISO 33004 suggest you should. Whether the project manager selected the right process, performs the practice, or does them in the right way is the reason for an audit or review and not that of a process capability assessment. Incapable people could, and do, use capable processes. Humans have always introduced random errors. But this is no different than the use of any capability assessment, such as COBIT™ 5 or TIPA (The Tudor IT Service Management Process Assessment). No IT office is the same as any other IT office. Organizations do not have identical technology. No organization offers the exact services as another organization. This uniqueness is what provides organizations short-term competitive advantage. But in none of these assessment methods does anyone proscribe certain actions and prescribe any set of processes. Your organization sets its goals and then focuses on those processes that align with them. Section 3.3 of ISO 21500 further states that "The processes selected for performing a project should be aligned in a systemic view." Therefore, our assessment methodology aligns with ISO 21500. Even when we provide a list of work products, assessors use their expertise to determine whether there is sufficient evidence that the process meets its purpose. These ideas dovetail with Agile methodologies, such as Crystal, Dynamic Systems Development Method (DSDM), Extreme Programming (XP), Rational Unified

Process (RUP) and Scrum. Tailoring is key to project management and the use of our assessment methodology.

Q5. What happens when there are not four instances for the assessment?
A5. If you have read through to this point, then you know a Class 1 assessment calls for a minimum of four instances, and a Class 2 calls for a minimum of two instances. This is an easy process when you are looking at Corrective Actions in processes CO04, CO06, CO07, CO08, CO09, CO10, and CO11 or Change Requests in processes PL15, IM02, IM04, IM05, CO01, CO03, CO04, CO06, CO07, CO08, CO09, and CO10. You may have a surfeit of change requests to assess. But when you look at something such as process ES02 and its output of a project management methodology, we would expect only one instance; that is, the project management methodology document itself. In truth, for something like this, should you should have four instances, you should have only one established project management methodology that everyone follows. For processes that do not have the minimum instances required, you assess all the instances available.

Q6. Do your assessors need certification?
A6. No, but your assessors do need subject matter expertise. However, it would, and might help, when your assessors have process knowledge and certification as a COBIT or TIPA assessor.

Q7. Who does the certification?
A7. Currently no one offers certification.

Q8. What are some quantitative risk assessment methodologies?
A8. IEC 31010[4] offers information about quantitative risk assessment methodologies such as failure mode effects analysis (FMEA), fault tree analysis (FTA), decision trees, bow-tie analysis, Markov analysis, Monte Carlo simulation, Bayesian statistics and Bayes Nets, and multi-criteria decision analysis.

PMI's *Practice Standard for Project Risk Management* (2009) lists decision tree analysis, expected monetary value (EMV), FMEA/FTA, and Monte Carlo simulation.

---

[4] Refer to IEC 31010:2009, *Risk management—Risk assessment techniques.*

Q9. Are there any ISO quality standards?
A9. The major one is *ISO 9001:2015, Quality management systems—Requirements*. But there are other specific standards such as *ISO 13053-1:2011, Quantitative methods in process improvement—Six Sigma—Part 1: DMAIC methodology*, ISO 13053-2:2011, *Quantitative methods in process improvement—Six Sigma—Part 2: Tools and techniques* and *ISO 18404:2015, Quantitative methods in process improvement—Six Sigma—Competencies for key personnel and their organizations in relation to Six Sigma and Lean implementation.*

Q10. Are there any ISO process improvement standards?
A10. Absolutely, every management system standard, such as ISO/IEC 27001, is a process improvement standard as they are based on ISO 9001. ISO 21500 is a process improvement standard, it is just not a management system. Perhaps in the future, ISO will have a project management system standard.

Q11. What are the key PM processes?
A11. There are no key project management processes. But processes in the Establish (ES) and Monitor (MO) domains are necessary to succeed. So, if this is your first project management assessment, we would recommend you focus your assessment on the Establish domain. After that, planning processes are important for successful projects.

Q12. How do I decide what percentage to assign as a rating?
A12. Well this is a little trickier question. As part of the assessment plan, you must specify how you will rate the processes. The simplest way is to calculate the percentage. For example, when there are 12 required artifacts and they have eight, then you give them an L, since this is 66.67 percent. If they have four, they get a P since this is 33.33 percent.

Alternatively, you could use your professional judgement. You could use some weighting schema based on the importance of the artefact. You could base the rating on the contribution of the artefacts to meeting the purpose. Think about the Pareto Principle: 20 percent of the practices and work products will provide 80 percent of the value. Hopefully, you appreciate that not all artefacts are created equal.

To us, to get an L or Largely, you just need to do the process in a perfunctory manner.

The key is to document the rating method and get agreement before starting the assessment on how you will calculate these numbers. As long as you could justify your rating method to a peer, and they could replicate it, it is satisfactory and quite acceptable.

Q13. Do I need to be a mathematician to do ratings?

A13. Good news for you innumerate, math-phobic, polynomial-panicked people out there. Everything you need to know to do a rating you learnt by grade 3. The toughest math we do is percentages and averages. If you are struggling with that, then get yourself an app or a 9-year-old.

Q14. What it is the difference between ordinal, interval and ratio scales?[5]

A14. With ordinal scales, it is the order of the values that is important and significant, but the differences between each one is not really known. An ordinal scale denotes direction. Take a look at the example below:

How satisfied are you with our service?

1. Very dissatisfied
2. Somewhat dissatisfied
3. Neither satisfied or dissatisfied
4. Somewhat satisfied
5. Very satisfied[6]

Intuitively, we know that a 4 is better than a 3 or 2, but we don't know–and cannot quantify–how much better it is. For example, is the difference between "Very Satisfied" and "Somewhat Satisfied" the same as the difference between "Very Dissatisfied" and "Somewhat Dissatisfied?" Who knows? We cannot even define satisfied for you. What satisfies you might dissatisfy us. Think of our levels, is the difference between Level 1 and Level 2 the same as the difference between Level 4 and Level 5? We think not.

---

[5] There is another scale: nominal. It deals more with quality then quantity. Binary 0 and 1 are nominal values, since they represent off and on.

[6] This is called a Likert scale.

Have you have ever gone to the hospital and the doctor asks, "on a scale of 1–10, how much pain do you have?" That is an ordinal value. When calculating central tendency using ordinal data, you can calculate the mode or median; but you cannot calculate the mean.

Interval scales are numeric scales where we know not only the order, but also the exact differences between the values. The classic example of an interval scale is Celsius temperature because the difference between each value is the same. For example, the difference between 20° and 30° is a measurable 10°, as is the difference between 10° and 20°. We have used interval scales for ratings.

However, think about those Fahrenheit and Celsius scales. We think you agree that 20°F is not twice as hot as 10°F. And the ten degrees between –10°C and 0°C seems a lot worse than the difference between 0°C and 10°C. This is because it is not a true ratio. When calculating central tendency using interval data, you can calculate the mode, median, or mean; and standard deviation. The real problem with an interval scale is there is no true zero. Think of our rating N which goes from 0% to 15%. How does one get a zero? The very fact that you are doing a project management assessment means you have some knowledge of project management, hence you are not an absolute zero! Without an absolute or true zero, you cannot calculate ratios.

Which brings us to ratio scales. Ratio scales tell us about the order, have an exact value between units, and have an absolute zero, which allows for a wide range of both descriptive and inferential statistics to be applied. Absolute or true zero is a point where none of the quality being measured exists. In keeping with the temperature theme, the Kelvin scale has no negative numbers and an absolute zero, where molecular motion stops.[7] Other good examples of ratio variables are height, weight and money. Too many know what zero money means.

To summarize:

To compensate for some of its shortcomings, people often treat ordinal values as though they were interval. They kludge the scale

---

[7] Again, zero degrees K is the point where the fundamental particles of nature have minimal vibrational motion, retaining only quantum mechanical, zero-point energy-induced particle motion.

Table A1.1 Data Types and Measures

MEASURE	ORDINAL	INTERVAL	RATIO
Order	✓	✓	✓
Counts or Frequency of Distribution	✓	✓	✓
Mode	✓	✓	✓
Median	✓	✓	✓
Percentiles	✓	✓	✓
Mean		✓	✓
Difference between quantifiable		✓	✓
Standard deviation, standard error		✓	✓
Add or subtract values		✓	✓
Multiple and divide			✓
Coefficient of variation			✓
True zero			✓

and make the difference between 1 and 2 the same as the difference between 2 and 3 (Table A1.1).

So, should you really want to do comparisons, stay away from nominal and ordinal scales.

Q15. What capability level should I aim to achieve?

A15. *This is a consultant's favorite question*: It is akin to asking us "how long is a piece of rope?" In both cases, the answer is "it depends." Your organization needs to set targets, and then assess your organization against those targets. Some processes support directly the organizations objectives, and you may want to achieve a higher rating for those processes. Perhaps, you want to be world-class in those processes. While others do not directly support the objectives and goals, so it is quite acceptable to be at Level 1. But as a rule of thumb when deciding which level to aim for, risk goes down as the level goes up. Even within a level, the risk goes up when the you are missing more artefacts or not performing as many practices as when the gap is slight. To illustrate, the following table provides some insight when picking the level and writing your final report (Table A1.2).

Obviously, the assessors' expertise helps them decide what are minor, significant and major gaps.

Table A1.2 Consequence and Risk of Gap

LEVEL	CONSEQUENCE BY LEVEL	RISK BY LEVEL AND GAP		
		MINOR NON-CONFORMANCE	SIGNIFICANT NON-CONFORMANCE	MAJOR NON-CONFORMANCE
1	Missing work products, process outcomes not achieved. Consequence very high.	Medium	High	High
2	Cost or time overruns or unpredictable product or service quality. Consequence high.	Medium	Medium	High
3	Inconsistent project management performance across the organization. Consequence medium.	Low	Medium	Medium
4	Inability to quantify performance or proactively detect process problems. Consequence low.	Low	Low	Medium
5	Inability to innovate, ideate, achieve or evaluate process improvements. Consequence very low.	Low	Low	Low

Q16. Why can I not rate achievement as Level 2.5 or 3.5?

A16. Well, you could, but you would not be compliant with ISO 33000 standards. You are either a two or a three. But we are not the "ISO police" and as we said, the standards are voluntary, so you could do anything you want to achieve your objective of improving your project management processes. Several organizations we know do in fact give themselves a value such as 2.3. They believe they show progress by then achieving 2.4. We believe you show progress by developing action plans and then providing status to management on their completion. But, hey, whatever floats your boat.

Appendix D: Terms and Definitions

This appendix provides definitions for terms used throughout the book.

TERMINOLOGY	DEFINITION
Activity	Identified component of work within a schedule that is required to be undertaken to complete a project (ISO 21500)
Assessment	The act of making a judgement about something, in our case, the Project Methodology and its components
Assessment indicator	An assessment indicator that supports the judgement of the extent of achievement of a specific process attribute (ISO 33000)
Assessment team	One or more individuals who jointly perform a process assessment (ISO 33000)
Assessor	Individual who participates in the rating of process attributed (ISO 33000)
Artifact	Something observed by the assessor created as a result of a process; such as policies, procedures, and practices
Attribute	A characteristic or part of something. Each attribute measures the particular specific or generic practices involved with each capability level
Attribute indicator	As assessment indicator that supports the judgement of the extent of achievement of a specific process attribute (ISO 15504-2)
Base practice	An activity that, when consistently performed, contributes to achieving a specific process purpose (ISO 33000)

(*Continued*)

TERMINOLOGY	DEFINITION
Basic maturity level	Lowest level of achievement in a scale of organizational process maturity (ISO 33000)
Basic process set	Set of processes that ensure the achievement of the basic maturity level (ISO 33000)
Capability dimension	The set of elements in a process assessment model explicitly related to the Measurement Framework for Process Capability (ISO 33000)
Capability indicator	An assessment indicator that supports the judgement of the process capability of a specific process (ISO 33000)
Change request	Documentation that defines a proposed alteration to the project (ISO 21500)
Configuration management	Application of procedures to control, correlate and maintain documentation, specifications, and physical attributes (ISO 21500)
Control	Comparison of actual performance with planned performance, analyzing variances, and taking appropriate corrective and preventive action as needed (ISO 21500)
Corrective action	Direction and activity for modifying the performance of work to bring performance in line with the plan (ISO 21500)
Defined process	A process that is managed (planned, monitored and adjusted), and tailored from the organization's set of standard processes according to the organization's tailoring guidelines
Generic practice	An activity that, when consistently performed, contributes to the achievement of a specific process attribute (ISO 33000)
Instance	A set of process activities or certain types of processes, for example, a single change record in change management for example
Lead assessor	Assessor who had demonstrated the competencies to conduct an assessment and to monitor and verify the conformance of a process assessment (ISO 33000)
Maturity level	Point of an ordinal scale of organizational process maturity that characterizes the maturity of the organizational unit assessed in the scope of the maturity model used (ISO 33000)
Objective evidence	Data supporting the existence or verity of something (ISO 33000)
Organizational process maturity	The extent to which an organizational unit consistently implements processes within a defined scope that contributes to the achievement of its business needs (current or projected) (ISO 33000)
Performance indicator	An assessment indicator that supports the judgement of the process performance of a specific process (ISO 33000)
Practice	An activity that contributes to the purpose or outcomes of a process or enhances the capability of a process

(*Continued*)

TERMINOLOGY	DEFINITION
Preventive action	Direction and activity for modifying the work in order to avoid or reduce deviations in performance from the plan (ISO 21500)
Process	Set of interrelated or interacting activities that transform inputs into outputs (ISO 9000)
Process assessment model	A model suitable for the purpose of assessing process capability, based on one or more process reference models (ISO 33000)
Process attribute	A measurable characteristic of process capability applicable to any process (ISO 33000)
Process attribute outcome	Observable result of achievement of a specified process attribute (ISO 33000)
Process attribute rating	A judgement of the degree of achievement of the process attribute for the assessed process (ISO 33000)
Process capability	Characterization of the ability of a process to meet current or projected goals (ISO 33020)
Process capability level	A point on the six-point ordinal scale (of process capability) that represents the capability of the process, each level building on the capability of the level below. Characterization of a process on an ordinal measurement scale of process capability (ISO 33020)
Process measurement framework	Schema for use in characterizing a process quality characteristic of an implemented process (ISO 33000)
Process outcome	An observable result of a process. An outcome is an artifact, a significant change of state or the meeting of specified constraints
Process purpose	The high-level measurable objectives of performing the process and the likely outcomes of effective implementation of the process (ISO 33000)
Process quality	Ability of a process to satisfy stated and implied stakeholder needs when used in a specified context (ISO 33000)
Process quality attribute	Measurable property of a process quality characteristic (ISO 33000)
Process quality characteristic	Measurable aspect of process quality; category of process attributes that are significant to process quality. Process quality characteristics include properties of processes such as process capability, efficiency, effectiveness, security, integrity and sustainability (ISO 33000)
Process quality dimension	Set of elements in a process assessment model explicitly related to the process measurement framework for the specified process quality characteristic (ISO 33000)
Process quality level	Point on a scale of achievement of a process characteristic derived from the process attribute ratings for an assessed process (ISO 33000)

(*Continued*)

TERMINOLOGY	DEFINITION
Process reference model	A model composed definitions of processes in a life cycle described in terms of process purpose and outcomes, together with an architecture describing the relationships between the processes (ISO 33000)
Project	Consists of a unique set of processes consisting of coordinated and controlled activities with start and end dates, performed to achieve project objectives (ISO 21500)
Project management	The discipline of initiating, planning, executing, controlling, and closing the work of a team to achieve specific goals and meet specific success criteria
Project management plan	The document that describes how the project will be executed, monitored and controlled (PMBOK)
Stakeholder	Person, group or organization that has interests in, or can affect, be affected by, or perceive itself to be affected by any aspect of the project (ISO 21500)
Variation	An actual condition that is different from the expected condition that is contained in the baseline plan (PMBOK)
Work breakdown structure	A hierarchical decomposition of the total scope of work to be carried out by the project team to accomplish the project objectives and create the required deliverables (PMBOK)
Work breakdown structure dictionary	Document that describes each component in the work breakdown structure (ISO 21500)
Work product	An artifact associated with the execution of a process (ISO 9000). Artifacts are both inputs and outputs to processes. There are four generic product categories, as follows: services (e.g., operation); software (e.g., computer program, documents, information, contents); hardware (e.g., computer, device); and processed materials. Project management itself is a service, but it may deliver software, hardware or processes materials

Appendix E: Acronyms and Initialisms

This appendix provides acronyms, such as ISACA, and initialisms, such as OPM3.

TERM	MEANING
ANSI	American National Standards Institute
APM	Association for Project Management
CMM	Capability Maturity Model
CL	Close domain
CMMI	Capability Maturity Model Integration
CMMI-DEV	CMMI® for Development
CO	Control domain
COBIT	Originally stood for Control Objectives for IT and related technology. Now used only as its Acronym
DSDM	Dynamic systems development method
EMV	Expected monetary value
eSCM-CL	eSourcing Capability Model for Client Organizations
eSCM-SP	eSourcing Capability Model for Service Providers
ES	Establish domain
FMEA	Failure mode effect analysis
FTA	Fault tree analysis
IEC	International Electrotechnical Commission
IM	Implement domain
IN	Initiate domain

(*Continued*)

TERM	MEANING
ISACA	Originally stood for Information Systems Audit and Control Association; now uses only its acronym
ISO	International Organization for Standardization
ITIL	Information Technology Infrastructure Library
MO	Monitor domain
OPM3	Organizational Project Management Maturity Model
PCA	Process capability assessment
PDTS	Preliminary draft technical specification
PL	Plan domain
PMBOK	Project Management Body of Knowledge
PMI	Project Management Institute
PMP	Project Management Professional
PRINCE2	Projects IN Controlled Environments 2
PRM	Process reference model
RUP	Rational Unified Process
SEI	Software Engineering Institute
SPI	Software Process Improvement
SPICE	Software Process Improvement and Capability dEtermination
SW-CMM	Capability Maturity Model for Software
SWEBOK	Software Engineering Body of Knowledge a.k.a. ISO/IEC TR 19759:2005
TIPA	Tudor IT Process Assessment
TR	Technical Report
TS	Technical Standard

Appendix F: References

1. AccountAbility. *AA1000 Stakeholder Engagement Standard.* AccountAbility Standards Board: 2015. Retrieved from https://www.accountability.org/wp-content/uploads/2016/10/AA1000SES_2015.pdf (retrieved date April 4, 2018).
2. Cabinet Office. *An Introduction to PRINCE2™: Managing and Directing Successful Projects: 2009 ed.* The Stationery Office (UK): 2009. Retrieved from https://www.axelos.com/store/book/managing-successful-projects-with-prince2 (retrieved date April 4, 2018).
3. Department of Energy/Department of Homeland Security. *Cybersecurity Capability Maturity Model (C2M2), Version 1.1.* US Government: February 2014. Retrieved from http://energy.gov/oe/services/cybersecurity/cybersecurity-capability-maturity-model-c2m2-program (retrieved date April 4, 2018).
4. International Organization for Standardization. *ISO/IEC 15504-4:2004, Information technology—Process assessment—Part 4: Guidance on use for process improvement and process capability determination.* ISO: 2012. Retrieved from https://www.iso.org/standard/37462.html (retrieved date April 4, 2018).
5. International Organization for Standardization. *ISO/IEC 15504-5:2012, Information technology—Process assessment—Part 5: An exemplar software life cycle process assessment model.* ISO: 2012. Retrieved from https://www.iso.org/standard/60555.html (retrieved date April 4, 2018).
6. International Organization for Standardization. *ISO/IEC 15504-6:2013, Information technology—Process assessment—Part 6: An exemplar system life cycle process assessment model.* ISO: 2013. Retrieved from https://www.iso.org/standard/61492.html (retrieved date April 4, 2018).

7. International Organization for Standardization. *ISO/IEC TS 15504-8:2012, Information technology—Process assessment—Part 8: An exemplar process assessment model for IT service management.* ISO: 2012. Retrieved from https://www.iso.org/standard/50625.html (retrieved date April 4, 2018).
8. International Organization for Standardization. *ISO/IEC TS 15504-9:2011, Information technology—Process assessment—Part 9: Target process profiles.* ISO: 2011. Retrieved from https://www.iso.org/standard/51684.html (retrieved date April 4, 2018).
9. International Organization for Standardization. *ISO/IEC 20000-1:2011, Information technology—Service management—Part 1: Service management system requirements.* ISO: 2011. Retrieved from https://www.iso.org/standard/51986.html (retrieved date April 4, 2018).
10. International Organization for Standardization. *ISO 21500:2012, Guidance on project management.* ISO: 2012. Retrieved from https://www.iso.org/standard/50003.html (retrieved date April 4, 2018).
11. International Organization for Standardization. *ISO/IEC 25010:2011, Systems and software engineering—Systems and software Quality Requirements and Evaluation (SQuaRE)—System and software quality models.* ISO: 2011. Retrieved from https://www.iso.org/standard/35733.html (retrieved date April 4, 2018).
12. International Organization for Standardization. *ISO/IEC 27001:2013, Information technology—Security techniques—Information security management systems—Requirements.* ISO: 2013. Retrieved from http://www.iso.org/iso/catalogue_detail?csnumber=54534 (retrieved date April 4, 2018).
13. International Organization for Standardization. *ISO/IEC TR 29110-3-1:2015, Systems and software engineering—Lifecycle profiles for Very Small Entities (VSEs)—Part 3-1: Assessment guide.* ISO: 2015. Retrieved from https://www.iso.org/standard/62713.html (retrieved date April 4, 2018).
14. International Organization for Standardization. *ISO/IEC 29169:2016, Information technology—Process assessment—The application of conformity assessment methodology to the assessment to process quality characteristics and organizational maturity.* Retrieved from https://www.iso.org/standard/45248.html (retrieved date April 4, 2018).
15. International Organization for Standardization. *ISO/IEC 33001:2015, Information technology—Process assessment—Concepts and terminology.* ISO: 2015. Retrieved from https://www.iso.org/standard/54175.html (retrieved date April 4, 2018).
16. International Organization for Standardization. *ISO/IEC 33002:2015, Information technology—Process assessment—Requirements for performing process assessment.* ISO: 2015. Retrieved from https://www.iso.org/standard/54176.html (retrieved date April 4, 2018).
17. International Organization for Standardization. *ISO/IEC 33003:2015, Information technology—Process assessment—Requirements for process measurement frameworks.* ISO: 2015. Retrieved from https://www.iso.org/standard/54177.html (retrieved date April 4, 2018).

18. International Organization for Standardization. *ISO/IEC 33004:2015, Information technology—Process assessment—Requirements for process reference, process assessment and maturity models.* ISO: 2015. Retrieved from https://www.iso.org/standard/54178.html (retrieved date April 4, 2018).
19. International Organization for Standardization. *ISO/IEC TR 33014:2013, Information technology—Process assessment—Guide for process improvement.* ISO: 2013. Retrieved from https://www.iso.org/standard/54186.html (retrieved date April 4, 2018).
20. International Organization for Standardization. *ISO/IEC 33020:2015, Information technology—Process assessment—Process measurement framework for assessment of process capability.* ISO: 2015. Retrieved from https://www.iso.org/standard/54195.html (retrieved date April 4, 2018).
21. International Organization for Standardization. *ISO/IEC TS 33052:2016, Information technology—Process reference model (PRM) for information security management.* ISO: 2016. Retrieved from https://www.iso.org/standard/55142.html (retrieved date April 4, 2018).
22. International Organization for Standardization. *ISO/IEC 33053 PDTS1 Information Technology—Process Assessment—Process reference model for quality management.* ISO: TBD.
23. International Organization for Standardization. *ISO/IEC 33063:2015, Information technology—Process assessment—Process assessment model for software testing.* ISO: 2015. Retrieved from https://www.iso.org/standard/55154.html (retrieved date April 4, 2018).
24. International Organization for Standardization. *ISO/IEC 33071:2016, Information technology—Process assessment—An integrated process capability assessment model for Enterprise processes.* ISO: 2016. Retrieved from https://www.iso.org/standard/55162.html (retrieved date April 4, 2018).
25. International Organization for Standardization. *ISO/IEC TS 33072:2016, Information technology—Process assessment—Process capability assessment model for information security management.* ISO: 2016. Retrieved from https://www.iso.org/standard/70803.html (retrieved date April 4, 2018).
26. International Organization for Standardization. *ISO/IEC 33073 PDTS1 Information technology—Process assessment—Process capability assessment model for Quality Management.* ISO: TBD. Retrieved from https://www.iso.org/standard/55164.html (retrieved date April 4, 2018).
27. International Organization for Standardization. *ISO/IEC 38500:2015, Information technology—Governance of IT for the organization.* ISO: 2015. Retrieved from http://www.iso.org/iso/home/store/catalogue_tc/catalogue_detail.htm?csnumber=62816 (retrieved date April 4, 2018).
28. International Organization for Standardization. *ISO/IEC 38502:2014, Information technology—Governance of IT—Framework and model.* ISO: 2014. Retrieved from http://www.iso.org/iso/home/store/catalogue_tc/catalogue_detail.htm?csnumber=50962 (retrieved date April 4, 2018).

29. ISACA. *COBIT 5: A Business Framework for the Governance and Management of Enterprise IT.* ISACA: 2012. Retrieved from http://www.isaca.org/COBIT/Pages/Product-Family.aspx (retrieved date April 4, 2018).
30. ISACA. *COBIT Assessor Guide: Using COBIT 5.* ISACA: 2012. Retrieved from http://www.isaca.org/COBIT/Pages/Assessor-Guide.aspx (retrieved date April 4, 2018).
31. ISACA. *COBIT Process Assessment Model (PAM): Using COBIT 5.* ISACA: 2012. Retrieved from http://www.isaca.org/COBIT/Pages/COBIT-5-PAM.aspx (retrieved date April 4, 2018).
32. ISACA. *COBIT Self-Assessment Guide: Using COBIT 5.* ISACA. Retrieved from http://www.isaca.org/COBIT/Pages/Self-Assessment-Guide.aspx (retrieved date April 4, 2018).
33. National Institute of Standards and Technology. *Risk Management Framework (RMF).* NIST: January 2017. Retrieved from http://csrc.nist.gov/groups/SMA/fisma/Risk-Management-Framework (retrieved date April 4, 2018).
34. Project Management Institute. *A Guide to the Project Management Body of Knowledge (PMBOK® Guide)—Sixth Edition.* PMI: 2017. Retrieved from https://www.pmi.org/pmbok-guide-standards/foundational/pmbok/sixth-edition (retrieved date April 4, 2018).
35. Project Management Institute. *Navigating Complexity, A Practice Guide.* PMI: Newtown Square, PA: 2014. Retrieved from http://www.pmi.org/pmbok-guide-standards/practice-guides/complexity (retrieved date April 4, 2018).
36. Project Management Institute. *Organizational Project Management Maturity Model (OPM3): Knowledge Foundation.* PMI: 2013. Retrieved from https://www.pmi.org/pmbok-guide-standards/foundational/organizational-pm-maturity-model-opm3-third-edition (retrieved date April 4, 2018).
37. Project Management Institute. *Practice Standard for Project Risk Management.* PMI: 2009. Retrieved from https://www.pmi.org/pmbok-guide-standards/practice-guides/complexity (retrieved date April 4, 2018).
38. Roebuck, E. Guide to process capability assessment and process control, *Production Engineer*, 52, (5), 165–170, 1973. doi: 10.1049/tpe.1973.0027.
39. URL: http://ieeexplore.ieee.org/stamp/stamp.jsp?tp=&arnumber=4914022&isnumber=4914019 (retrieved date April 4, 2018).
40. Software Engineering Institute. *CMMI for Development, Version 1.3.* SEI: 2010. Retrieved from http://resources.sei.cmu.edu/library/asset-view.cfm?assetID=9661 (retrieved date April 4, 2018).
41. Software Engineering Institute. *CMMI® for Services, Version 1.2.* SEI: 2009. Retrieved from http://www.sei.cmu.edu/reports/09tr001.pdf (retrieved date April 4, 2018).

42. Tudor, D. J. *Agile Project and Service Management: Delivering IT Services Using PRINCE2*. The Stationery Office: London, UK, July 28, 2010.
43. Valdés, O. *ITSM Process Assessment Supporting ITIL (TIPA)*. Van Haren Publishing: 2009. Retrieved from https://www.vanharen.net/shop/itsm-process-assessment-supporting-itil-tipa?__store=vanharen_net_en&__from_store=vanharen_net_nl (retrieved date April 4, 2018).

Appendix G: Assessor Guide Checklist

ASSESSOR GUIDE CHECKLIST			
TASK	DESCRIPTION	ACCOMPLISHED?	DATE
1	Determine the business need		
2	Determine the class of assessment		
3	Get sponsor approval		
4	Determine the category of independence needed		
5	Set out roles and responsibilities		
6	Assess competency requirements		
7	Document the class of assessment and category of independence		
8	Determine the assessment scope		
9	Determine communications to the staff involved		
10	Set cut the activities to be performed		
11	Assign the resources to be used		
12	Determine resource schedules		
13	Document assessment inputs		
14	Describe assessment outputs		
15	Decide whether to initiate a pre-assessment questionnaire		
16	Describe the strategy and techniques for the selecting, identifying, collecting and analyzing objective evidence and data		
17	Document sponsor approval of the plan		
18	Set up key interviews		
19	Determine record traceability plan		

(*Continued*)

ASSESSOR GUIDE CHECKLIST			
TASK	DESCRIPTION	ACCOMPLISHED?	DATE
20	Review collected data		
21	Follow evidence requirements		
22	Apply traceability		
23	Determine whether the evidence is objective		
24	Determine whether the evidence is sufficient		
25	Decide whether it is representative enough to cover the assessment purpose and class		
26	Determine whether the evidence is consistent as a whole		
27	Ensure that the defined set of assessment indicators are used		
28	Select rating and aggregation methods		
29	Apply traceability		
30	Verify traceability of evidence		
31	Decide if the ordinal scale will be further refined for the measures P and L		
32	Apply ratings		
33	Write report		

Appendix H: Sample Data Tracking Form

When performing an assessment, this form may provide a method for tracking each work product item for compliance with traceability requirements.

Example

PROCESS ID		PROCESS NAME		
OUTCOMES	WORK PRODUCTS REVIEWED	ITEM TRACKING NO	VERIFIED?	COMMENTS
ES01-01	*INPUTS*			Use ES01-IN1a, b, and c for multiple documents/interview notes
A project management framework is defined	Decision-making model	ES01-IN1	Y	
	Authority limits	ES01-IN2	Y	
	Enterprise governance guiding principles	ES01-IN3	Y	
ES01-02	Process architecture model	ES01-IN4	N	
A project management framework is followed				

(*Continued*)

PROCESS ID	PROCESS NAME			
OUTCOMES	WORK PRODUCTS REVIEWED	ITEM TRACKING NO	VERIFIED?	COMMENTS
	OUTPUTS			
	Project management guidelines	ES01-OUT1	Y	
	PMO Charter	ES01-OUT2	Y	
	Definition of organizational structure and functions	ES01-OUT3	N	
	Definition of project-related roles and responsibilities	ES01-OUT4	Y	
	Process capability assessments	ES01-OUT5	Y	
	Performance goals and metrics of process improvement tracking	ES01-OUT6	Y	

Appendix I: Process Ranking Form

Complete each column as follows:

- Rank strategic importance on a scale from 1 (not important) to 5 (very important).
- Rank process performance on a scale from 1 (done well) to 5 (not done well).
- Rank degree of formality on a scale from 1 (very formal and well-documented) to 5 (very information and not documented).
- Estimate date of last assessment from 1 (current year) to 5 (five or more years since last audit).
- Multiply columns 3–6 to derive rank priority number. Work on the processes with the highest numbers first. Do not use this number for anything other than establishing rank—it is based on a Likert scale.

PROCESS ID (1)	PROCESS (2)	STRATEGIC IMPORTANCE (3)	PROCESS PERFORMANCE (4)	DEGREE OF FORMALITY (5)	LAST AUDIT (6)	RANK PRIORITY NUMBER (3 * 4 * 5 * 6)
ESTABLISH DOMAIN						
ES01	Define the project management framework					
ES02	Set policies, processes and methodologies					
ES03	Set limits of authority for decision-making					
MONITOR DOMAIN						
MO01	Ensure project benefits					
MO02	Ensure risk optimization					
MO03	Ensure resource optimization					
INITIATE DOMAIN						
IN01	Develop project charter					
IN02	Identify stakeholders					
IN03	Establish project team					
PLAN DOMAIN						
PL01	Develop project plans					
PL02	Define scope					
PL03	Create work breakdown structure					
PL04	Define activities					
PL05	Estimate resources					
PL06	Define project organization					
PL07	Sequence activities					
PL08	Estimate activity durations					
PL09	Develop schedule					

(*Continued*)

PROCESS ID (1)	PROCESS (2)	STRATEGIC IMPORTANCE (3)	PROCESS PERFORMANCE (4)	DEGREE OF FORMALITY (5)	LAST AUDIT (6)	RANK PRIORITY NUMBER (3 * 4 * 5 * 6)
PL10	Estimate costs					
PL11	Develop budget					
PL12	Identify risks					
PL13	Assess risks					
PL14	Plan quality					
PL15	Plan procurements					
PL16	Plan communications					
CONTROL DOMAIN						
CO01	Control project work					
CO02	Control changes					
CO03	Control scope					
CO04	Control resources					
CO05	Manage project team					
CO06	Control schedule					
CO07	Control costs					
CO08	Control risks					
CO09	Perform quality control					
CO10	Administer procurements					
CO11	Manage communications					
CLOSE DOMAIN						
CL01	Close project phase or project					
CL02	Collect lessons learned					

Appendix J: Key Steps in an Assessment

1. Determine the business need
2. Determine the class of assessment
3. Get sponsor approval
4. Determine the category of independence needed
5. Establish roles and responsibilities
6. Assess competency requirements
7. Document the class of assessment and category of independence
8. Determine the assessment scope
9. Determine communications to the staff involved
10. Set out the activities to be performed
11. Assign the resources to be used
12. Determine resource schedules
13. Document assessment inputs
14. Describe assessment outputs
15. Decide whether to initiate a pre-assessment questionnaire
16. Describe the strategy and techniques for the selecting, identifying, collecting and analyzing objective evidence and data
17. Document sponsor approval of the plan
18. Set up key interviews
19. Determine record traceability plan
20. Review collected data

21. Follow evidence requirements
22. Determine whether the evidence is objective
23. Determine whether the evidence is sufficient
24. Decide whether it is representative enough to cover the assessment purpose and class
25. Determine whether the evidence is consistent as a whole
26. Ensure that the defined set of assessment indicators are used
27. Select rating and aggregation methods
28. Verify traceability of evidence
29. Decide if the ordinal scale will be further refined for the measures P and L
30. Write report
31. Ensure recommended Table of Contents is followed

Index

Note: Page numbers followed by f and t refer to figures and tables respectively.

S

T

V

W

9781138298521